MONOGRAPHS ON
APPLIED PROBABILITY AND STATISTICS

General Editors

M.S. BARTLETT, F.R.S., *and* D.R. COX, F.R.S.

STOCHASTIC ABUNDANCE MODELS

Stochastic Abundance Models

WITH EMPHASIS ON BIOLOGICAL COMMUNITIES
AND SPECIES DIVERSITY

S. ENGEN

Department of Mathematics
University of Trondheim

LONDON
CHAPMAN AND HALL

A Halsted Press Book

John Wiley & Sons, New York

First published 1978
by Chapman and Hall Ltd.
11 New Fetter Lane, London EC4P 4EE

© 1978 S. Engen

Photosetting by Thomson Press (India) Limited, New Delhi
and printed in Great Britain by
J.W. Arrowsmith Ltd., Bristol 3

ISBN 0 412 15240 1

Distributed in the U.S.A. by Halsted Press,
a Division of John Wiley & Sons, Inc., New York

Library of Congress Cataloging in Publication Data

Engen, S
Stochastic abundance models, with emphasis on
biological communities and species diversity.

(Monographs on applied probability and statistics)
Based on the author's thesis, Oxford.
"A Halsted Press book."
Bibliography: p.
Includes indexes.
1. Animal populations—Statistical methods.
2. Stochastic processes. I. Title.
QH352.E53 1978   574.5′24′0182   77-28215
ISBN 0-470-26302-4

Contents

Preface

This monograph deals with the analysis of populations of elements. Each element is a member of one and only one class, and we shall mainly be concerned with populations with a large number of classes. No doubt the present theory has its outspring in ecology, where the elements symbolize the individual animals or plants, while the classes are the various species of the ecological community under consideration. Some basic ideas point back to a classical contribution by R.A. Fisher (1943, in collaboration with A.S. Corbet and C.B. Williams) representing a breakthrough for the theoretical analysis of diverse populations. Though most of the work in this field has been carried out by ecologists, statisticians and biometricians have, over the past 15 years, shown an ever increasing interest in the topic.

Besides being directed towards biometricians and statisticians, this monograph may hopefully be of interest for any research worker dealing with the classification of units into a large number of classes, in particular ecologists, sociologists and linguists. However, some background in statistics and probability theory is required. It would be unless to read the present book without some knowledge of the continuous and discrete probability distributions summarized in section 1.1, and the use of generating functions. In particular, a clear intuitive and formal understanding of the concept of conditional probability and conditional distributions is required in order to interpret the various models correctly.

The analysis of diverse populations is naturally divided into three parts: (1) Description of populations of classes, abundance models, and parameter estimation; (ii) generating processes; (iii) environmental influence on the diversity (of classes) and the role of interaction between classes. This monograph deals with (i) only. Generating processes will not be dealt with, except for some comments on the subject at the end of Chapter 3. We may consider (iii) as a further development of (i). In fact, one main purpose of this text is to present a theory that can form a basis for a study of factors controlling the diversity. However, it would be too large an area to discuss here, nor would the present author be the right person to do so.

The simultaneous description of more than one population is not dealt with. Such models, including the measurement of similarity between populations, are as yet rather poorly developed. It is, however, an important field, and its problems represent a challenge to research workers in biometry, ecology and the other branches of science mentioned above.

As regards the material included in the monograph, Part I, the general theory, deals with the basic description of populations of classes, abundance models, and the estimation of population parameters. We also examine possible deductions relating to the 'species–area' curve and the concept of sample coverage; equations that establish relations between model parameters and generally applicable indices of diversity and equitability will be derived in Chapter 5. Part I is written without reference to particular applications, and it is hoped the presentation may prove useful to any scientist faced with populations consisting of a large number of classes.

Part II is devoted to ecology. We here discuss the realism of the various assumptions upon which different types of models are based. Through examples of the analysis of real biological data we show how to assess the applicability of some of the models presented in Part I. The typical ecological (botanical) sampling scheme known as 'quadrat sampling' is mentioned briefly in Part I and dealt with in greater detail in Part II.

Mathematics Department,   Steinar Engen
University of Trondheim
September 1977

Acknowledgments

I gratefully acknowledge the help and advice of a number of friends and scholars in the preparation of this book. T. Strømgren in particular deserves my thanks for arousing my interest in diversity problems and for the encouragement and stimulation he has provided over the years. I am also glad to take this opportunity of expressing my gratitude to Professor A. Høyland, who willingly backed up all my applications for financial support.

The present monograph is based on my D. Phil. thesis *Statistical Analysis of Species Diversity*. Professor M.S. Bartlett and Dr M.G. Bulmer supervised my work for the thesis, at the Department of Biomathematics, University of Oxford, during the period 1972–74. I am greatly indebted to both for their sound advice and penetrating criticism at all stages. The research carried out in connection with my thesis was supported by grants from the Norwegian Research Council for Science and the Humanities in collaboration with The Royal Society, as part of the European Science Exchange Programme.

Further, I wish to acknowledge my debt to Eva Seim, who read parts of the draft and suggested a number of improvements in the presentation; to H. Waadeland for valuable suggestions concerning the proof in Appendix A; to S. Svensson, who kindly placed some of his Christmas data at my disposal; to Professor S. Haftorn, who established this contact; and to K. Aagaard, who helped me with the example on Chironomid communities. The tables in Appendix B are reproduced by kind permission from *Biometrics*.

I. Siegismund typed the manuscript of the present book. I am indebted to her for her patient and meticulous work, much of it undertaken at short notice and under great pressure. Finally, I would like to thank L.E. Breivik for reading the draft in its entirety and for checking my English.

S.E.

We must attend to the quantitative aspect of a situation or problem and make a basic quantitative analysis. Every quality manifests itself in a certain quantity, and without quantity there can be no quality. To this day many of our comrades still do not understand that they must attend to the quantitative aspect of things—the basic statistics, the main percentages and the quantitative limits that determine the qualities of things. They have no figures in their heads and as a result cannot help making mistakes.

Mao Tse-Tung: 'Methods of Work of
Party Committees' (13 March, 1949)

PART I

Theoretical treatment

CHAPTER ONE

Introduction

1.1 Abundance models

In this monograph we shall use the term 'abundance model' in a more general sense than that which is commonly used in ecological literature. There can be little doubt that ecology was responsible for the development of this theory and is still the main field of application. Though we shall mention other fields as well, the applications shown in Part II will be mainly ecological. However, in view of the possibility of expansion into other branches of science and also into the arts, I have found it convenient to separate the theory from its applications. In ecology, an 'abundance model' is a mathematical/statistical model describing features of an animal- (or plant-) population of many species relating to the relative abundances. In the general theory, we shall let the population be a set of elements (individuals) where each element belongs to one and only one of the classes $C_1, C_2, \ldots, C_s$. Since both finite and infinite populations will be dealt with, there is a possibility that $s = \infty$ as well. Let p_i be the proportion of elements belonging to $C_i, i = 1, 2, \ldots, s$, and write $\mathbf{p} = (p_1, p_2, \ldots, p_s)$. The general aim of an abundance model is to make a study of $\mathbf{p}$ in cases where s is large. We are not interested in one particular p_i but in the whole vector $\mathbf{p}$. This may be some real valued functions of $\mathbf{p}$, such as the information

index of diversity, $H = -\sum_{j=1}^{s} p_j \ln p_j$ (Shannon, 1948), but it may

also be a law stating some sort of statistical regularity in (or between) the components of $\mathbf{p}$. In either case there will be parameters to draw inferences about and, furthermore, a need for statistical treatment of data obtained from random samples of elements. Tests for changes in parameter values will also be required.

A further, and quite important step would be to investigate how the population† $\mathbf{p}$ is generated. This would imply considering the stochastic process $\mathbf{p}(t)$, where t is time, and search for realistic processes in agreement with the observed regularity. The study of such processes is beyond the scope of this book, though we shall touch on the subject in Chapter 3.

1.2. Some distributions used in the statistical analysis of abundance and diversity

Notation

Let $\mathbf{X} = (X_1, X_2, \ldots, X_s)$ be a stochastic vector. If $\mathbf{X}$ is distributed continuously, we shall let $f_\mathbf{X}(\mathbf{x})$ denote the density of $\mathbf{X}$; and if $\mathbf{X}$ is discrete, $P_\mathbf{X}(\mathbf{x})$ will denote the point density. The probability generating function will be abbreviated p.g.f. and symbolized by

$$G_\mathbf{X}(\mathbf{z}) = E(z_1^{X_1} z_2^{X_2} \ldots z_s^{X_s}).$$

The moment generating function is abbreviated m.g.f. and symbolized by

$$M_\mathbf{X}(\mathbf{t}) = E(e^{t_1 X_1 + t_2 X_2 + \ldots + t_s X_s}).$$

The following is a list of distributions along with some of their properties, stated without proofs. Readers who want proofs of these statements, which will be referred to in different contexts throughout this book, may consult a work on general probability theory and statistics (e.g. Kendall & Stuart, 1969).

(a) *The gama distribution* (*type III distribution*) The gamma density with parameters (α, κ) is

$$\frac{\alpha^\kappa}{\Gamma(\kappa)} x^{\kappa-1} e^{-\alpha x}, \qquad \alpha, \kappa > 0, \qquad x \geqslant 0.$$

Here $\alpha > 0$ is a trivial scale parameter, while κ is the essential shape parameter. The expectation equals α/κ and the variance α/κ^2. The rth moment is $\Gamma(\kappa + r)/\alpha^r \Gamma(\kappa)$. The gamma densities are closed under convolution; that is, if X_1 and X_2 are independent gamma variates

---

† We shall use the term 'population' for a collection of elements that is subdivided into classes. This is consistent with standard statistical literature. However, in the ecological application where the elements are individuals and the classes are species, the biological term 'community' would be more adequate. We shall use the latter term in Part II.

with parameters (α, κ_1) and (α, κ_2), then $X_1 + X_2$ is a gamma variate with parameters $(\alpha, \kappa_1 + \kappa_2)$.

(b) *The beta distribution (of the first kind)* The beta density with parameters (p, q) is

$$\frac{1}{B(p,q)} x^{p-1}(1-x)^{q-1}, \qquad p, q > 0, \qquad 0 \leqslant x \leqslant 1,$$

where

$$B(p,q) = \frac{\Gamma(p)\Gamma(q)}{\Gamma(p+q)}.$$

The expectation is $p/(p+q)$, the variance $pq/[(p+q)^2(p+q+1)]$, and the rth moment $\Gamma(p+r)\Gamma(p+q)/[\Gamma(p)\Gamma(p+q+r)]$. If X_1 and X_2 are independently gamma distributed with parameters (α, κ_1) and (α, κ_2) respectively, then $X_1/(X_1 + X_2)$ is beta distributed with parameters (κ_1, κ_2). On the basis of the moment sequence it is easy to show that if Y_1 and Y_2 are independently beta distributed with parameters (p_1, q_1) and $(p_1 + q_1, q_2)$, then their product $Y_1 Y_2$ is also beta distributed with parameters $(p_1, q_1 + q_2)$.

(c) *The beta distribution of the second kind* The beta density of the second kind with parameters (p, q) is

$$\frac{1}{B(p,q)} \frac{x^{p-1}}{(1+x)^{p+q}}, \qquad p, q > 0, \qquad x \geqslant 0.$$

If X possesses the beta distribution of the first kind with parameters (p, q), then $X/(1+X)$ possesses the beta distribution of the second kind with parameters (p, q). The rth moment is $\Gamma(p+r)\Gamma(q-r)/[\Gamma(p)\Gamma(q)]$ for $r < q$ and does not exist for $r \geqslant q$.

(d) *The lognormal distribution* If X is normally distributed with mean μ and variance σ^2, we shall say that $Y = e^X$ is lognormally distributed with parameters (μ, σ^2). The moments are

$$E(Y^r) = Ee^{rX} = e^{r\mu + (1/2)r^2\sigma^2}, r = 0, 1, 2, \ldots$$

Hence, in particular

$$E(Y) = e^{\mu + (1/2)\sigma^2}, \operatorname{var} Y = e^{2\mu + \sigma^2}(e^{\sigma^2} - 1).$$

(See Aitchison and Brown, 1969).

(e) *The Dirichlet distribution* Let $X_1, X_2, \ldots, X_s$ be independent gamma variates with parameters $(\alpha, \kappa_1), (\alpha, \kappa_2), \ldots, (\alpha, \kappa_s)$. Write $P_i = X_i \Big/ \sum\limits_{j=1}^{s} X_j$. Then the simultaneous density of $P_1, P_2, \ldots, P_s$ is the Dirichlet density

$$f_p(\mathbf{p}) = \Gamma\left(\sum_{i=1}^{s} \kappa_i\right) \cdot \prod_{i=1}^{s} \left[\frac{p_i^{\kappa_i - 1}}{\Gamma(\kappa_i)}\right], \qquad \Sigma p_i = 1.$$

The marginal distribution of p_j is the beta distribution of the first kind with parameters $(\kappa_j, \sum\limits_{i \ne j} \kappa_i)$.

(f) *The negative binomial distribution* For a given $\lambda > 0$ let X be a Poisson variate with mean λ. Then, if λ is gamma distributed with parameters (α, κ) the unconditional distribution of X is

$$P_X(x) = \beta^\kappa \frac{\Gamma(\kappa + x)}{\Gamma(\kappa) x!} (1 - \beta)^x, \qquad x = 0, 1, 2, \ldots$$

where $\beta = \alpha/(\alpha + 1)$. This is the negative binomial distribution with parameters (β, κ). The p.g.f. is $[1 - (1 - \beta)z]^{-\kappa}$, giving $E(X) = k/\alpha$ and $\mathrm{var}(X) = \dfrac{\kappa}{\alpha} + \dfrac{\kappa}{\alpha^2}$.

(g) *The logarithmic series distribution* Omitting the zero class in the negative binomial, we get

$$P_X(x|X > 0) = \frac{\beta^\kappa}{1 - \beta^\kappa} \frac{\Gamma(\kappa + x)}{\Gamma(\kappa) x!} (1 - \beta)^x, \qquad x = 1, 2, \ldots$$

(sometimes called the zero-truncated negative binomial and is in fact well defined for any $\kappa > -1$ (Engen, 1974)). The limit of this distribution as $\kappa \to 0$ is $[1/-\ln \beta][(1 - \beta)^x/x]$, which is called the logarithmic series distribution. The corresponding limit of the p.g.f. is $\ln[1 - (1 - \beta)z]/\ln \beta$, giving

$$E(X) = \frac{\delta(1 - \beta)}{\beta}, \quad \mathrm{var}\, X = \frac{\delta(1 - \beta)[1 - \delta(1 - \beta)]}{\beta^2},$$

where $\delta = (-\ln \beta)^{-1}$.

(h) *The negative binomial beta distribution* Let X be a negative binomial variate with parameters (β, κ) and write $\beta = \alpha/(\alpha + 1)$

as in (f). Now, considering the mixture on α, where α has the beta distribution of the second kind with parameters (p, q), we arrive at the distribution

$$P_X(x) = \frac{\Gamma(\kappa + p)\Gamma(p + q)}{\Gamma(\kappa)\Gamma(p)\Gamma(q)} \frac{\Gamma(\kappa + x)\Gamma(q + x)}{\Gamma(x + 1)\Gamma(\kappa + p + q + x)}, \quad x = 0, 1, 2, \ldots$$

We shall call this the negative binomial beta distribution with parameters (κ, p, q). (Also called the generalized hypergeometric distribution type IV; see Patil and Joshi, 1968). Alternatively, we may consider the mixture on β with the beta distribution of the first kind with parameters (p, q).

(i) *The Poisson–lognormal distribution* The mixture on λ of the Poisson distribution with mean λ with the lognormal distribution with parameters (μ, σ^2) will be referred to as the Poisson–lognormal distribution with parameters (μ, σ^2). The distribution is given by the integral

$$P_X(x) = \frac{e^{x\mu + (1/2)x^2\sigma^2}(2\pi\sigma^2)^{-1/2}}{x!} \int_{-\infty}^{\infty} e^{-e^y - (y - \mu - x\sigma^2)^2/2\sigma^2} dy$$

The rth factorial moment of this distribution is

$$\mu_{[r]} = e^{r\mu + (1/2)r^2\sigma^2}, r = 1, 2, \ldots \qquad \text{(Bulmer, 1974)}.$$

1.3. Description of fixed populations

1.3.1. *Introduction*

In this section we shall consider populations at a particular fixed point in time. Hence, they are not subject to any form of randomness or stochastic variation. This may be a useful approach, even if some population features are quite obviously of a stochastic nature. In fact, one is often most interested in what the population was like at the time of sampling and not what it might have been if the random elements of nature had resulted in a different realization. In any case, results based on the assumption of a fixed population may be considered as conditional upon the state of the population at the time of sampling, and therefore as preliminary to the more general approach.

1.3.2. *Finite populations*

Now consider a population of n elements, where each element belongs to one and only one of the classes C_1, C_2,..., C_s, where n and s are both finite numbers. We shall mainly be interested in populations with a relatively large number of classes. Let x_i be the number of elements in C_i, $i = 1, 2, ..., s$, and write r_j, where $j = 1, 2, ..., n$, for the number of classes in which there are exactly j elements. Hence $s = \sum_{j=1}^{n} r_j$ and $n = \sum_{j=1}^{n} jr_j$. We shall say that the vector $\mathbf{r} = (r_1, r_2, ..., r_n, 0, 0, ...)$ is the class structure of the population. Two populations are said to possess the same class structure if and only if they are associated with the same vector $\mathbf{r}$.

Example 1.1. In the ecological context, the classes may correspond to some taxonomic classification of animals or plants. The elements may be the individual organisms of some taxonomic group living in a particular area, say, an island or a lake, or some area bounded artificially by the person carrying out the investigation. The corresponding classes would be the various species of the taxonomic group under consideration. Alternatively, species may serve as elements and genera as classes; higher order classifications can also be considered.

Example 1.2. In linguistics, the elements may be, say, the grammatical units, or morphemes, of a text; the classes may be various groups of equal morphemes. What is to be understood by the term *equal morphemes* has of course to be defined in advance by the linguist. As far as the English language is concerned, the class structure of a text depends on whether, say, *was* and *were* are defined as members of the same class or not.

Example 1.3. The following example has also been taken from English linguistics. This time the elements are all the sounds occurring in a spoken utterance, and the classes are the various groups of distinctive, or meaning-differentiating sounds (i.e. the phonemes). The class structure (or phonemic inventory) depends on whether, say, the diphthong *au* in *house* is interpreted as one phoneme or whether it is split up and assigned to the phonemes *a* and *u* respectively. (Linguists do, in fact, differ on this point).

Example 1.4. In a sociological study, the elements may be the individual people of some community, and the classes may be the various occupations represented in the community.

Example 1.5. The elements may be the individual items within a certain group of items on sale in a community. The classes may be the groups of items produced in factories owned by the same company.

Example 1.6. Good (1953) considered chess openings in games published in the British Chess Magazine, 1951. The opening of two games was regarded as equivalent if the first six moves (three white and three black) were the same and in the same order in both games. The example, which is mainly of academic interest, fits into our theory if we let the games be the elements and view the openings as classes.

A feature common to the above examples is that the class structure is likely to reveal some characteristic population properties that are of interest to the research worker dealing with it. When the class structure, or certain features of the structure, has been established, there still remain some relevant questions to ask (though perhaps not for example 1.6.); for example, what made the structure become what it is, and why do different populations sometimes show quite different structures? In attempting to answer such questions, it is necessary to develop statistical methods that enable us to draw inferences about the population structure. In practice it will be necessary to concentrate on some functions of r, and we shall use the term population parameters for such functions. By way of analogy, consider the following example.

Example 1.7. Let $y_1, y_2, \ldots, y_n$ be the heights of the adult male population of a city. We define the population mean and variance by

$$\bar{y} = \frac{1}{n}\Sigma y_i, \quad \sigma_y^2 = \frac{1}{n-1}\Sigma(y_i - \bar{y})^2.$$

These are population parameters expressing the central tendency and the variability respectively, and they alone give us some idea of what the whole vector y is like.

For populations with a class structure the population parameters most commonly used are the so-called indices of diversity, which are supposed to have at least one of the following properties.

(1) They increase with s.
(2) They increase as the components of $\mathbf{x}$ tend to be 'more equal' in value.

For finite populations, the diversity measure most commonly used apart from s itself is the Brillouin combinatorial type of measure

$$H_b = n^{-1} \ln\left(\frac{n!}{x_1!\,x_2!\ldots x_s!}\right) \tag{1.1}$$

(Brillouin, 1962). The above index has been taken from information theory, but is now often used by ecologists. Alternatively H_b may be written in the form

$$H_b = n^{-1} \sum_{i=1}^{n} \left[\ln i - r_i \ln(i!)\right]. \tag{1.2}$$

Another parameter often used is the Simpson index of diversity (Simpson, 1949)

$$H_s = 1 - \sum_{i=1}^{s} \frac{x_i(x_i - 1)}{n(n-1)} = 1 - \sum_{i=1}^{n} \frac{r_i i(i-1)}{n(n-1)}. \tag{1.3}$$

H_s is the probability that two elements chosen at random from the population belong to different classes. Indices of this type will be discussed in further detail in Chapter 5.

1.3.3. *Infinite populations*

In practice the number of elements in the population is quite large compared with the size of the samples we are able to take. As n is usually unknown, it is convenient to consider idealized populations with an infinite number of elements. Write $p_i = x_i/n$ for the proportion of elements belonging to C_i. Then, given $\mathbf{p} = (p_1, p_2, \ldots, p_s)$ and n, the population structure is uniquely defined. In the limit $n \to \infty$, $\mathbf{p}$ is still well defined with $\sum_{i=1}^{s} p_i = 1$. Note that, in passing on to the limit, we open up the possibility of letting $s \to \infty$ as well. Some models with an infinite number of classes will be dealt with in Chapter 3. The proportion p_i will be called the relative abundance of C_i.

Authors dealing with the concept of diversity sometimes also operate in terms of absolute measures of abundance. For a finite population this is x_i itself, but for infinite populations absolute abundances are most conveniently defined by some method of sampling, say, as the expected number of elements of C_i appearing in a sample obtained by some prescribed method of sampling. For infinite populations the absolute abundance of C_i will be denoted λ_i, and we shall in general assume that $\lambda_i = vp_i$. Hence

$$v = \sum_{i=1}^{s} \lambda_i$$ is the expected number of elements in the sample and

is therefore naturally called the sample size. In particular, the relative abundance p_i of C_i is the expected number of elements of C_i appearing in a sample of unit size.

1.3.4. Description of structure by distributions

Let us assume that s is finite. Let us further suppose that one class is selected at random from the list of classes in such a way that each class has the same probability of being chosen. Let Λ and P be the absolute and relative abundance of this class respectively. Then Λ and P possess the discrete distributions (Holgate, 1969).

$$Pr(\Lambda = \lambda_i) = Pr(P = p_i) = \frac{1}{s}, i = 1, 2, \dots, s. \tag{1.4}$$

We shall write $g(\lambda | \lambda)$ for the above distribution assigning the probability measure $1/s$ to each point λ_i, and similarly $h(p|\mathbf{p})$ for the distribution of P.

Now, these distributions uniquely define the values of the components of λ and $\mathbf{p}$, but they are quite useless in the case $s = \infty$, which is illustrated by the following theorem.

Theorem 1.1. Suppose that $g(\lambda | \lambda)$ and $h(p|\mathbf{p})$ are defined as above. Let s be the number of positive components of λ and $\mathbf{p}$ and assume that $v = \Sigma \lambda_i$ is given. Then $g(\lambda | \lambda)$ and $h(p|\mathbf{p})$ do not converge to distributions as $s \to \infty$.

Proof. Suppose, without loss of generality, that $p_i > 0$ for $i = 1, 2, \dots, s$, and $p_i = 0$ for $i > s$. Let l_ε be the number of components of $\mathbf{p}$ with the property $p_i > \varepsilon$. Since $\Sigma p_i = 1$, $\varepsilon l_\varepsilon \leqslant 1$ or $l_\varepsilon \leqslant 1/\varepsilon$. Hence

$$Pr(P \leqslant \varepsilon) = \frac{s - l_\varepsilon}{s} \geqslant 1 - \frac{1}{\varepsilon s},$$

implying that $\lim_{s \to \infty} Pr(P \leqslant \varepsilon) = 1$ for any $\varepsilon > 0$. If the distribution of P exists in the limit, it must therefore be entirely concentrated at zero. But $Pr(P > 0) = 1$ by definition, and hence, the limiting distribution does not exist. The same holds for $g(\lambda | \lambda)$, since $\Lambda = vP$.

Much of the controversy about Fisher's logarithmic series model (Fisher, Corbet and Williams, 1943), which we shall deal with in Chapter 3, is linked to the content of Theorem 1.1. Since we shall be working with models with an infinite number of classes, it will be convenient to introduce distributions associated with populations defined in such a way that the limiting distributions are also well-defined. Instead of selecting the classes with equal probabilities, we now assume that C_i is chosen with probability p_i. Let Λ^* and P^* be the absolute and relative abundance of the class chosen in this manner. Hence

$$Pr(\Lambda^* = \lambda_i) = Pr(P^* = p_i) = p_i, \qquad i = 1, 2, \dots, s. \qquad (1.5)$$

and write shortly $g^*(\lambda | \lambda)$ and $h^*(p | \mathbf{p})$ for these distributions. $g^*(\lambda | \lambda)$ and $h^*(p | \mathbf{p})$ are clearly well-defined in the limit $s = \infty$, since $\sum_{i=1}^{\infty} p_i = 1$. We shall refer to $h^*(p | \mathbf{p})$ as the *structural distribution*† associated with the population.

1.4. Description of random populations

The number of elements in C_i at a particular point of time is constant, but x_i and $\mathbf{p}$ will in general be functions of time. In fact, $\mathbf{p}(t)$ is a multivariate stochastic process that would usually be quite intractable. This is because the number of components are quite large and interact in a complicated way. Some difficulties in describing such processes in ecology will be deferred until Chapter 6, but whether or not a successful description of the above time series can be given, we may define some random experiment resulting in a point of time T, which is then a random variable. Hence, $\mathbf{x}(T)$, $\mathbf{r}(T)$ and $\mathbf{p}(T)$ are also random variables. More generally, we may be interested in a set of populations, and consider the random experiment of selecting one population at random. In linquistics the populations may be various texts, in sociology different cities,

---

† Our terms 'structure' and 'structural distribution' refer to the ecological concept 'structure of a community'. It should not be confused with the term 'structural inferences' used in statistical literature (Fraser, 1968)

in ecology perhaps the populations of different locations. If we are, in this manner, going to consider randomized populations (i.e. treating for example **p** as a random variable), it is extremely important for the applicability of the methods explored that the assumptions made about the distribution of **p** are realistic. We therefore leave the discussion of particular stochastic models for **p** to Part II. At this stage, we shall only generalize the description by means of the random variables introduced in the previous section. Suppose that Λ, P and P^* conditionally upon **p** are defined as above. Let $f(\mathbf{p})$ be the density of **p** and define

$$h^*(\mathbf{p}) = \int h^*(p\,|\,\mathbf{p})f(\mathbf{p})d\mathbf{p} \qquad (1.6)$$

and let $h(p)$ and $g(\lambda)$ be defined in the same way. Hence it appears that $h^*(p)$ is the unconditional distribution of P^*. We shall use the term 'structural distribution' for $h^*(p)$ as well. Note that neither **p** nor $f(\mathbf{p})$ can be expressed by $h^*(p)$ in this case, but we shall see in the next chapter that one sample from the population will mainly reveal properties of $h^*(p)$ rather than the simultaneous density of **p**. In cases where it is appropriate to work with absolute abundances we define

$$g^*(\lambda) = \frac{\lambda g(\lambda)}{\int \lambda g(\lambda)d\lambda} \qquad (1.7)$$

for the case that $g(\lambda) = Eg(\lambda|\lambda)$ exists (we shall let Λ^* denote a random variable with density $g^*(\lambda)$ in this case).

For sequences of populations for which s tends to infinity we have seen that $g(\lambda)$ does not tend to a limiting density. However, the limit of $g^*(\lambda)$ will exist, and we shall later find it useful.

Example 1.8. Let **p** possess the Dirichlet distribution

$$f(\mathbf{p}) = \frac{\Gamma(\kappa s)}{\Gamma(\kappa)^s}(p_1 p_2 \ldots p_s)^{\kappa-1}, \qquad \Sigma p_i = 1, \qquad \kappa > 0.$$

Consider the moment sequence

$$E(P^{*r}\,|\,\mathbf{p}) = \sum_{i=1}^{s} p_i^{r+1}, r = 0, 1, 2, \ldots$$

Unconditionally, the moment sequence is

$$E(P^{*r}) = \int \left(\sum_{i=1}^{s} p_i^{r+1} \right) f(\mathbf{p})d\mathbf{p} = \frac{\Gamma(\kappa s + 1)\Gamma(\kappa + r + 1)}{\Gamma(\kappa + 1)\Gamma(\kappa s + r + 1)},$$

which is in fact the moment sequence for

$$h^*(p) = \frac{\Gamma(\kappa s + 1)}{\Gamma(\kappa + 1)\Gamma(\kappa s - \kappa)} p^\kappa (1 - p)^{\kappa s - \kappa - 1},$$

which is the beta-distribution with parameters $\kappa + 1$ and $\kappa(s - 1)$.

Example 1.9. Let $v(p)$ be some function and write

$$V(\mathbf{p}) = \sum_{i=1}^{s} v(p_i).$$

Then

$$V(\mathbf{p}) = \sum_{i=1}^{s} \frac{1}{p_i} v(p_i) p_i = E\left[\frac{1}{P^*} v(P^*) \Big| \mathbf{p}\right]$$

Consequently

$$E[V(\mathbf{p})] = \int \frac{1}{p} v(p) h^*(p) dp.$$

In particular if $v(p) = 1$ we obtain

$$V(\mathbf{p}) = E[V(\mathbf{p})] = s = \int \frac{1}{p} h^*(p) dp.$$

Example 1.10. Write $f(\mathbf{p})$ for the density of $\mathbf{p}$, and $f_i(p)$ for the marginal density of $p_i, i = 1, 2, \ldots, s$. The structural distribution for a fixed model with s classes may be written

$$h^*(p \,|\, \mathbf{p}) = \sum_{i=1}^{s} p_i \delta(p - p_i),$$

where $\delta(\cdot)$ is Dirac's δ-function. Hence we have unconditionally

$$h^*(p) = \int \sum_{i=1}^{s} p_i \delta(p_i - p) f(\mathbf{p}) d\mathbf{p} = \int \sum_{i=1}^{s} p_i \delta(p_i - p) f_i(p_i) dp_i.$$

As a result we get the relation

$$h^*(p) = p \sum_{i=1}^{s} f_i(p).$$

By the same technique, writing

$$h(p \,|\, \mathbf{p}) = \frac{1}{s} \sum_{i=1}^{s} \delta(p_i - p)$$

we get

$$h(p) = \frac{1}{s} \sum_{i=1}^{s} f_i(p).$$

Hence in general for $s < \infty$

$$h^*(p) = sph(p).$$

Further, if $f(\mathbf{p})$ is symmetric, then $f_1(p) = f_2(p) = \ldots = f_s(p)$, giving $h(p) = f_i(p)$ and $h^*(p) = spf_i(p)$.

Example 1.11. By definition

$$g(\lambda \mid \lambda) = \frac{1}{s} \sum_{i=1}^{s} \delta(\lambda - \lambda_i).$$

$$\Sigma \lambda_i = s \sum \frac{1}{s} \lambda_i = s \int \lambda g(\lambda \mid \lambda) d\lambda \quad \text{and} \quad \mu = s \int \lambda g(\lambda) d\lambda,$$

where μ denotes E $(\Sigma \lambda_i) = E(v)$.
Let $w(\lambda)$ be some function of λ. Then

$$\Sigma w(\lambda_i) = s \int w(\lambda) g(\lambda \mid \lambda) d\lambda \quad \text{and} \quad E[\Sigma w(\lambda_i)] = s \int w(\lambda) g(\lambda) d\lambda$$

which by the above result may be written

$$E[\Sigma w(\lambda_i)] = \mu \int \frac{1}{\lambda} w(\lambda) g^*(\lambda) d\lambda$$

or

$$E[\Sigma w(\lambda_i)] = \mu E\left[\frac{w(\Lambda^*)}{\Lambda^*}\right].$$

Example 1.12. Let $f(\lambda)$ be the simultaneous density of λ and write $f_i(\lambda)$ for the marginal density of λ_i. Then

$$g(\lambda) = \int \frac{1}{s} \Sigma \delta(\lambda - \lambda_i) f(\lambda) d\lambda = \frac{1}{s} \Sigma f_i(\lambda).$$

Consequently, by definition

$$g^*(\lambda) = \frac{\lambda \Sigma f_i(\lambda)}{\int \lambda \Sigma f_i(\lambda) d\lambda}.$$

CHAPTER TWO

Sampling from a population of classes

2.1. Introduction

In order to demonstrate the difference between the sampling situation in the analysis of diversity and more classical sampling schemes, let us first reconsider Example 1.5., the heights of the adult male population in a city. The population mean and variance are defined by

$$\bar{y} = \frac{1}{n} \Sigma y_i \quad \text{and} \quad \sigma^2 = \frac{1}{n-1} \Sigma (y_i - \bar{y})^2.$$

In a random sample from this population, let the observed heights be $Y_1, Y_2, \ldots, Y_N$. Then

$$\bar{Y} = \frac{1}{N} \Sigma Y_i \quad \text{and} \quad s^2 = \frac{1}{N-1} \Sigma (Y_i - \bar{Y})^2$$

are unbiased estimators of $\bar{y}$ and σ^2 respectively.

When a random sample is drawn from a population of classes, the problems relating to estimation are usually much more complicated than in the above example. Let $\mathbf{x} = (x_1, x_2, \ldots, x_s)$ denote the number of elements in the various classes as in Chapter 1. If we were able to draw a random sample of classes and count the number of elements contained in each of them, then the situation would be analogous to the example with heights of people. But this would be a very unrealistic sampling situation, since in practice we must sample elements, not classes. If each element has the same chance of occurring in the sample, then, conditionally upon the number of elements sampled, $(X_1, X_2, \ldots, X_s)$ possesses the multidimensional hypergeometric distribution

$$P_{\mathbf{X}}(t_1, t_2, \ldots, t_s) = \frac{\binom{x_1}{t_1}\binom{x_2}{t_2}\cdots\binom{x_s}{t_s}}{\binom{n}{N}} \tag{2.1}$$

where $t_i = 0, 1, \ldots, \min(N, x_i)$ and $\Sigma t_i = N$
for a finite population, and the multinomial distribution

$$P_{\mathbf{X}}(t_1, t_2, \ldots, t_s) = \frac{N!}{t_1! t_2! \ldots t_s!} (p_1^{t_1} p_2^{t_2} \ldots p_s^{t_s}) \qquad (2.2)$$

where $t_i = 0, 1, \ldots, N$ and $\Sigma t_i = N$,
in the case $n = \infty$. But even these distributions (2.1 and 2.2) cannot form the basis of our analysis. This is due to the fact that we shall consider situations where nothing is known in advance about which classes are possibly represented in the population, nor do we know how many classes there are. Hence, although we shall assume that **X** possesses one of the above distributions when the abundances are considered as fixed, **X** is an unobservable variable*. But if our definition of the classes $C_1, C_2, \ldots, C_s$ is to be of any use at all, we must at least assume that an element is readily identifiable when it is examined; that is, by examining two elements, it is possible to determine whether they belong to the same class or not. This implies that the vector $\mathbf{R} = (R_1, R_2, \ldots, R_N)$ is observable, where R_i is the number of classes represented by exactly i elements in the sample. Since **R** is a function of **X**, the distribution of **R** for finite and infinite populations can in principle be derived from (2.1) and (2.2). Unfortunately, these distributions turn out to be rather intractable, as demonstrated below by a simple example.

Example 2.1. Put $s = N = 3$ in (2.2). Then the distribution of **R** is given by

$$P_{\mathbf{R}}(3, 0, 0) = 6 p_1 p_2 p_3$$

$$P_{\mathbf{R}}(1, 1, 0) = 3(p_1^2 p_2 + p_1^2 p_3 + p_1 p_2^2 + p_2^2 p_3 + p_1 p_3^2 + p_2 p_3^2).$$

If s is not specified we have for $N = 3$

$$P_{\mathbf{R}}(3, 0, 0) = 6 \sum_{i_j \neq i_k} p_{i_1} p_{i_2} p_{i_3},$$

$$P_{\mathbf{R}}(1, 1, 0) = 3 \sum_{i_1 \neq i_2} p_{i_1}^2 p_{i_2},$$

$$P_{\mathbf{R}}(0, 0, 1) = \sum_{i=1}^{s} p_i^3.$$

---

* The estimation of indices of diversity and equitability in the case of a multinomial model where $(X_1, X_2, \ldots, X_s)$ is actually observable is simpler to deal with than the case where only $(R_1, R_2, \ldots)$ is known. For the estimation of the information index for a multinomial model we refer to Basharin (1959) and Hutcheson (1970, 1974).

In general the distribution of $\mathbf{R}$ is given by

$$P_{\mathbf{R}}(t_1, t_2, \ldots, t_N) = N! \left(\prod_{i=1}^{N} (i!)^{t_i} \right)^{-1} \sum \left[\prod_{l=1}^{N} \left(\prod_{j=T_l+1}^{T_{l+1}} p_{i_j}^l \right) \right], \qquad (2.3)$$

where

$$T_1 = 0, \; T_l = \sum_{i=1}^{l-1} t_i, \; \sum_{i=1}^{n} i t_i = N,$$

and the sum is taken over all possible ways of selecting the t_1 subscripts in the first factor of the first product, the t_2 subscripts in the second factor, ..., the t_n subscripts in the nth factor. Consequently there are $S!(t_0! \, t_1! \ldots t_n!)^{-1}$ terms in the sum, where

$$t_0 = s - \sum_{i=1}^{n} t_i.$$

In the limit $s \to \infty$ (and $t_o \to \infty$) the number of terms is asymptotically of order $s^s (t_1! \, t_2! \ldots t_n!)^{-1}$.

It appears that the distribution of $\mathbf{R}$, as given by (2.3), is quite intractable and cannot be of much help in the search for inference methods. The situation is far more complicated than the classical multinomial case, where the set of classes in the population is known in advance.

The analogue of (2.3) for a finite population would be even more intractable.

2.2. The limitation of inference drawn from one sample

We define the contribution to R_j from the class C_i by

$$\Delta R_j^{(i)} = \begin{cases} 1 \text{ if } X_i = j \\ 0 \text{ otherwise.} \end{cases}$$

Then

$$R_j = \sum_i \Delta R_j^{(i)}.$$

For a given population λ, let the distribution of X_i be

$$P(X_i = j) = P(\Delta R_j^{(i)} = 1) = u_j(\lambda_i, \theta). \qquad (2.4)$$

Since $\lambda_i = E(X_i)$ we have

$$\sum_{j=0}^{\infty} j u_j(\lambda_i, \theta) = \lambda_i.$$

The parameter θ is an additional one, relevant to the actual sampling situation. If u_j is a one-parameter distribution, for example the Poisson distribution, then θ will not occur in the expression. Now

$$E(R_j \mid \lambda) = \sum_{i=1}^{s} E\Delta R_j^{(i)} = \sum_{i=1}^{s} u_j(\lambda_i, \theta).$$

Hence

$$E(R_j \mid \lambda) = s \sum_{i=1}^{s} u_i\left(\lambda_i, \theta\right)\frac{1}{s} = sE[u_j(\Delta, \theta)]$$

and

$$E(R_j) = sE[u_j(\Lambda, \theta)], \qquad (2.5)$$

where Λ is defined as in Chapter 1, and θ is some parameter relevant to the sampling procedure.

Example 2.2. Let the sampling be given by the Poisson distribution,

$$u_j(\lambda) = \frac{\lambda^j}{j!} e^{-\lambda}, \qquad j = 0, 1, 2, \ldots$$

Since the mean is the only parameter, θ does not occur in the distribution of X_i. Assume that

$$g(\lambda) = \frac{\rho^\kappa}{\Gamma(\kappa)} \lambda^{\kappa-1} e^{-\rho\lambda}$$

From (2.5) we have

$$E(R_j) = s \int_0^\infty \frac{\lambda^j}{j!} e^{-\lambda} g(\lambda) d\lambda = s\left(\frac{\rho}{\rho+1}\right)^\kappa \frac{\Gamma(\kappa+j)}{\Gamma(\kappa)j!} \left(\frac{1}{\rho+1}\right)^j,$$

$$j = 0, 1, 2, \ldots, \qquad (2.6)$$

showing that $E(R_j)$ is proportional to the terms of the negative binomial series.

Since $E(R_j)$ depends on the population structure only by way of the distribution of Λ, two populations that are different in structure but with identically distributed Λ would give the same sequence $E(R_1), E(R_2), \ldots$.

We thus see that estimation methods based on considering expected values of the components of **R** cannot give any information about parameters in the distribution of **p**, other than those appearing in the distribution of Λ.

Writing as in Example 1.11.

$$s \int \lambda g(\lambda) d\lambda = \mu = E(\Sigma \lambda_i),$$

equation (2.5) may be written [†]

$$E(R_j) = \frac{\mu}{\int \lambda g(\lambda) d\lambda} \cdot \int u_j(\lambda, \boldsymbol{\theta}) g(\lambda) d\lambda$$

$$= \mu \int \frac{u_j(\lambda, \boldsymbol{\theta})}{\lambda} g^*(\lambda) d\lambda,$$

or just

$$E(R_j) = \mu E\left[\frac{u_j(\Lambda^*, \boldsymbol{\theta})}{\Lambda^*}\right], \qquad j = 0, 1, 2, \ldots \qquad (2.7)$$

An important implication of (2.7) is that if u_j is the Poisson distribution, there is a one-to-one correspondence between the sequence $[E(R_j)]_{j=1}^{\infty}$ and the absolute structural distribution $g^*(\lambda)$. [‡] Conditional upon $v = \Sigma \lambda_i$, we have similarly

$$E(R_j \mid v) = \Sigma u_j(v P^*, \boldsymbol{\theta}) = E\left[\frac{u_j(v P^*, \boldsymbol{\theta})}{P^*}\right], \qquad j = 0, 1, \ldots \quad (2.8)$$

Again, if u_j is the Poisson distribution, there is a one-to-one correspondence between the sequence $\{E(R_j \mid v)\}_{j=1}^{\infty}$ and $h^*(p)$.

Example 2.3. (Engen, 1975a) In Example 1.8. we considered the Dirichlet distribution

$$f(\mathbf{p}) = \frac{\Gamma(\kappa s)}{\Gamma(\kappa)^s} (p_1 p_2 \ldots p_s)^{\kappa - 1}, \qquad \Sigma p_i = 1, \kappa > 0,$$

and derived the corresponding moment sequence $E(P^{*j}), j = 0, 1, 2, \ldots,$ and $h^*(p)$. Considering the limit $\kappa \to 0, s \to \infty$, so that $\kappa s \to \alpha > 0$, then we have

$$h^*(p) \to \alpha(1 - p)^{\alpha - 1}.$$

[†] Note that $\mu = E(v)$ when $v = \Sigma \lambda_i$ is considered a random variable.
[‡] In fact, the one-to-one correspondence hold if $g^*(\lambda)$ is continuous and $\int e^{-\lambda} [g^* \times (\lambda)]^2 \, d\lambda$ exists (Tricomi, 1955). I owe thanks to P.R. Andenaes for pointing this out to me.

We now turn to a very different model leading to the same distribution for P^*. Let $z_1, z_2, \ldots$ be a sequence of independent random variables with the distribution $\alpha(1-z)^{\alpha-1}$. Consider the sequence

$$q_1 = z_1, \qquad q_i = z_i \prod_{j=1}^{i-1} (1-z_j) \qquad i = 2, 3, \ldots$$

If $\gamma = \alpha/(\alpha+1)$, then the expectations $E(q_i)$ form the geometric series $\gamma(1-\gamma)^{i-1}, i = 1, 2, \ldots$. Since $q_1 + q_2 + \ldots + q_j \leqslant 1$ and $\lim_{j \to \infty} E(q_1 + q_2 + \ldots + q_j) = 1, q_1 + q_2 + \ldots + q_j$ tends to 1 in probability, and the q_i therefore may denote relative abundances. Note that for the above Dirichlet model the expected relative abundances are all s^{-1}; hence they tend to zero in the limit $s \to \infty$. Considering the moment sequence for P^* for the randomized geometric series defined above, we find

$$E(P^{*r}) = \sum_{i=1}^{\infty} E(q_i^{r+1}) = E(z_1^{r+1}) + \sum_{i=2}^{\infty} E(z_i^{r+1}) \prod_{j=1}^{i-1} E[(1-z_j)^{r+1}].$$

$$E(z_i^{r+1}) = \frac{\Gamma(\alpha+1)\Gamma(r+2)}{\Gamma(\alpha+r+2)} = a$$

$$E[(1-z_i)^{r+1}] = \frac{\alpha}{\alpha+r+1} = b$$

then

$$E(P^{*r}) = a \sum_{i=0}^{\infty} b^i = \frac{a}{1-b} = \frac{\Gamma(\alpha+1)\Gamma(r+1)}{\Gamma(\alpha+r+1)},$$

which is the same as for the above limit of the Dirichlet model.

Now, if the X_is, conditionally upon $\mathbf{p}$, are independent Poisson variates with means $p_i \nu$, then the sequences $[E(R_j|\nu)]_{j=1}^{\infty}$ will be equivalent for these two models.

It is also possible to construct quite different models that lead to *identical distributions* for $\mathbf{R}$ (Engen, 1977b). Consider a population with a finite number of classes and abundances $\lambda = (\lambda_1, \lambda_2, \ldots, \lambda_s)$, and write $f_\lambda(\lambda_1, \lambda_2, \ldots, \lambda_s)$ for the density of λ. We now define a set of populations, that is, a set of distributions f_λ, the members of which are indistinguishable when only one sample from the population is available, or in other words, the distribution of $\mathbf{R}$ is the same for any population element of this set. Let $\mathbf{J} = (j_1, j_2, \ldots, j_s)$ be the integers $1, 2, \ldots, s$ in some order and consider the transformation $\lambda' = \Phi(\lambda, \mathbf{J})$, or equivalently $\lambda'_i = \lambda_{j_i}, i = 1, 2, \ldots, s$. If $\mathbf{J}$ is constant, the class structure of the population λ' is equivalent to that of λ, since the classes are just renumbered. Writing $\mathbf{R}'$ for the analogue of $\mathbf{R}$ for the transformed population, then $\mathbf{R}' = \mathbf{R}$. The distribution of $\mathbf{R}'$ therefore does not depend on $\mathbf{J}$, and the distribution of $\mathbf{R}'$ is of course the same as that of $\mathbf{R}$. If we now let $\mathbf{J}$ be a random variable whose distribution depends on λ, say $Pr[\mathbf{J} = (i_1, i_2, \ldots, i_s) = \mathbf{I}] = P_{\mathbf{J}}(\mathbf{I}; \lambda)$, then $f_{\lambda'}$ may differ considerably from f_λ, while the distribution of $\mathbf{R}'$ is unchanged. This is so because $P_{\mathbf{R}'}(\mathbf{r} \mid \lambda) = \sum_{\mathbf{I}} P_{\mathbf{R}}(\mathbf{r} \mid \mathbf{I}, \lambda) P_{\mathbf{J}}(\mathbf{I}; \lambda) = P_{\mathbf{R}}(\mathbf{r} \mid \lambda)$ and therefore $P_{\mathbf{R}'}(\mathbf{r}) = P_{\mathbf{R}}(\mathbf{r})$.

Example 2.4. Let $\lambda_1, \lambda_2, \ldots, \lambda_s$ be independent gamma variates with parameters $(\alpha/v, \kappa)$, where $\alpha = \kappa s$. Hence, $\mathbf{P}$ possesses the Dirichlet distribution

$$f_{\mathbf{P}}(\mathbf{p}) = \frac{\Gamma(\kappa s)}{\Gamma(\kappa)^s} (p_1 p_2 \cdots p_s)^{\kappa - 1}, \sum_{i=1}^{s} p_i = 1 \ldots$$

Now let Σp_i be symbolized by a stick of unit length and let the p_i be mutually exclusive and exhaustive segments. We shall define $P_{\mathbf{J}}(\mathbf{I}; \lambda)$ by the following process: Let a point be chosen randomly on the line and let q_1 be the length of the segment in which this point is contained; q_1 is then identical with one of the p_i, say $q_1 = p_{j_1}$. Then, choosing another point on the remaining part of the line, we obtain $q_2 = p_{j_2}$. The process carried out by induction gives the whole set $q_r = p_{j_r}, r = 1, 2, \ldots, s$. It appears that the distribution of $\mathbf{J} = (j_1, j_2, \ldots, j_s)$ now depends only on $(p_1, p_2, \ldots, p_s)$, which is a function of λ. Writing $\lambda'_i = \lambda_{j_i}$, the population λ' is indistinguishable from λ.

In order to obtain more detailed information about this model, we define

$$Q_1 = q_1, \, Q_i = q_i \Big/ \left(1 - \sum_{j=1}^{i-1} q_j\right), i = 2, 3, \ldots, s.$$

It is now fairly straightforward to show, by induction, that P_1, $P_2, \ldots, P_s$ are independent beta variates, the distribution of Q_i being the beta distribution with parameters $(\kappa + 1, \alpha - \kappa i)$, for $i = 1, 2, \ldots, (s-1)$, while $Pr(P_s = 1) = 1$.

In the limit $\kappa \to 0, s \to \infty$ so that $\kappa s = \alpha$, the Q_i tends to be beta distributed with the same parameters $(1, \alpha)$. Hence, the limiting model is equivalent to that considered in Example 2.3.

We shall see in Chapter 3 that this transformed model is also well defined when $-1 < \kappa < 0$, though our starting point, the Dirichlet distribution, does not exist for $\kappa \leq 0$. It follows from Example 2.6. that the structural distribution of the transformed population is beta with parameters $(\kappa + 1, \alpha - \kappa)$ when $\kappa > 0$, and it will be shown in Appendix A that this holds for any $\kappa > -1$.

2.3. Sampling from a finite population

We shall now embark on the problem of estimating $r_j, j = 1, 2, \ldots, n$, for the finite population. This problem was considered by Goodman (1951) who showed that there exists one and only one unbiased estimator of the number of different classes $s = \Sigma r_j$, subject to the condition that the sample size is not less than the maximum number of elements in one class, that is, $q = \max(x_j) \leq N$. If this condition is not satisfied, no unbiased estimator exists. Write

$$a^{(i)} = \begin{cases} a(a - 1)\ldots(a - i + 1) & \text{for } i > 0 \\ 1 & \text{for } i = 0. \end{cases}$$

Goodman's estimator of s is

$$\hat{s} = \sum_{i=1}^{n} A_i R_i,$$

where

$$A_i = 1 - (-1)\frac{i(n - N + i - 1)^{(i)}}{N^{(i)}}$$

Goodman also gave a set of linear equations for an unbiased estimator of $(r_1, r_2, \ldots, r_q)$. Write

$$Pr(i|j, n, N) = \frac{\binom{j}{i}\binom{n - i}{N - i}}{\binom{n}{N}}$$

Then $\hat{r}_j, j = 1, 2, \ldots, q$, is the solution of

$$R_i = \sum_{j=1}^{n} Pr(i|j, n, N)\hat{r}_j, \quad i = 1, 2, \ldots, n. \tag{2.9}$$

The solution $\hat{r}$, is unbiased if $q \leq N$, otherwise there is no unbiased estimator of $(r_1, r_2, \ldots, r_q)$. From Goodman's formulation it would appear that the existence of an unbiased estimator of r_k depends on whether $q \leqslant N$ or not. It will be clear from Theorem 2.1. that this is not quite so. In fact, there exists always one and only one unbiased estimator of r_k if $k \leqslant N$, otherwise no unbiased estimator exists.

Theorem 2.1. Suppose we have s classes of n similar elements with x_1 elements in class 1, x_2 elements in class 2, ..., x_s elements in class s. The class of an element is readily identifiable when the element is examined. Let us suppose that a random sample of N elements is drawn without replacement. If R_k is the number of classes containing k elements in the sample and r_k is the number of classes containing k elements in the population, then there exists one and only one unbiased estimator

$$\hat{r}_k = \sum_{i=k}^{N} \left[\sum_{j=k}^{i} \frac{\binom{n}{j}\binom{i}{j}\binom{j}{k}}{\binom{N}{j}} (-1)^{j-k} \right] R_i \tag{2.10}$$

of r_k if $k \leqslant N = \Sigma i R_i$. There is no unbiased estimator of r_k if $k > N$.

Lemma. Let T be a statistic so that $E(T) = 0$. Then $T = 0$ at any sample point.

Proof of lemma. To prove the lemma, let the points in the sample space be ordered as proposed by Goodman (1951).

The sample space is ordered by increasing R_N; for equal values of R_N, order the sample points by increasing R_{N-1}; for equal values of R_{N-1} order the sample points by increasing values of R_{N-1}; ...; for equal values of R_3 order the points by increasing R_2. Let $O_i = [R_1(i), R_2(i), \ldots, R_N(i)]$ denote the ordered samples. To each O_i let us associate the population P_i with $r_1 = R_1 + n - N, r_i = R_i$ for $i = 2, 3, \ldots, N$. Then, as pointed out by Goodman, the only possible samples to be drawn from P_i are $O_1, O_2, \ldots, O_i$. Write $T(O_i)$ for the value of the estimator at O_i. Then if P_1 is the population, O_1 is the only possible sample and $T(O_1) = 0$ in order that $E(T) = 0$. If the population is P_2, then the possible samples are O_1 and O_2, $T(O_1)Pr(O_1|P_2) + T(O_2)Pr(O_2|P_2) = 0$ if $E(T) = 0$, hence $T(O_2) = 0$. The proof is completed by induction.

Proof of theorem 2.1. The proof of the theorem implies solving (2.9.).

However, it seems preferable to start off with a different set of linear equations. Let $X_i, i = 1, 2, \ldots, s$, be the number of elements of the ith class appearing in the sample. Then

$$E\left[\frac{X_i^{(k)}}{N^{(k)}}\right] = \frac{x_i^{(k)}}{n^{(k)}}, \quad k \leqslant N,$$

This follows from the fact that X_i possesses the hypergeometric distribution. By summation over all classes we obtain

$$a_k = \frac{\binom{n}{k}}{\binom{N}{k}} \sum_{i=1}^{N} i_{(k)} E(R_i) = \sum_{i=1}^{N} i_{(k)} r_i, \qquad k = 1, 2, \ldots, N.$$

It is a simple matter to verify that the solution of this equation is

$$r_k = \frac{1}{k!} \sum_{j=0}^{N-k} \frac{1}{j!} a_{k+j}(-1)^j, \qquad k = 1, 2, \ldots, N.$$

Hence, the unbiased estimator of r_k is

$$\hat{r}_k = \frac{1}{k!} \sum_{j=0}^{N-k} \frac{1}{j!} (-1)^j \frac{\binom{n}{k+j}}{\binom{N}{k+j}} \sum_{i=k}^{N} i_{(k)} r_i, \qquad k = 1, 2, \ldots, N.$$

This form is perhaps most suitable for numerical calculations; changing the order of summation, we arrive at (2.10). Since $E(\hat{r}_k) = E(\hat{\hat{r}}_k) = r_k, E(\hat{r}_k - \hat{\hat{r}}_k) = 0$, and hence by the lemma, $\hat{r}_k = \hat{\hat{r}}_k$; this is in fact the only unbiased estimator of r_k if $k \leqslant N$. It remains to be shown that no unbiased estimator of r_k exists if $k > N$. Let O_i and P_i be defined as in the proof of the lemma and suppose that $k > N$. For the population $P_1, r_k = 0$, hence $\hat{r}_k(O_1) = 0$ in order that $\hat{r}_k$ should be unbiased, and by induction, $\hat{r}_k(O_i)$ must be zero for any sample point. Consequently, $\hat{r}_k$ is not unbiased if $r_k > 0$.

2.4. Sampling from an infinite population

2.4.1. *Conditioning upon the number of elements sampled*

Conditionally upon N and $\mathbf{p}$ we assumed that $(X_1, X_2, \ldots, X_s)$

was a multinomial variate with probability generating function

$$G_X(\mathbf{z}|N) = \left(\sum_{i=1}^{s} p_i z_i \right)^N. \tag{2.11}$$

N is often not a number that can be chosen by the experimenter, but subject to random variations depending on the method of sampling. However, since the structure to be analysed is $\mathbf{p}$, it is unnecessary to consider the unconditional distribution. Knowledge of the distribution of N does not reveal any properties of $\mathbf{p}$ other than those that can be extracted from the conditional model. Intuitively, as pointed out by Fisher (1956, IV. 4) in a different context, N alone does not contain any information about $\mathbf{p}$, but the value of N determines the precision with which conclusions about $\mathbf{p}$ can be drawn. It is therefore appropriate to argue conditionally upon its observed value to ensure that we attach to the conclusion the precision actually achieved. We are not really interested in the precision obtained hypothetically in some situation that has in fact not occurred, though it possibly may be realized. However, it is sometimes convenient to assume that N is a Poisson variate with mean v. Again, this must be considered as a conditional model, now conditioning upon v. Then N possesses in reality a mixture of Poisson distributions. Conditioning upon v rather than N is mathematically convenient, since the components of $\mathbf{X}$ are then independent Poisson variates, the p.g.f. of $\mathbf{X}$ being

$$G_X(\mathbf{z}|v) = \prod_{i=1}^{s} e^{v p_i (z_i - 1)}. \tag{2.12}$$

The distributional properties of estimators, for example estimators of indices of diversity, are usually approximately the same, whether the conditioning variable is N or v.

Examples 2.5. and 2.6. are examples of distributions of N that may be realistic in some applications.

Example 2.5. Let N be a negative binomial variate with mean v and shape parameter κ. Then

$$G_N(z) = \left(\frac{\kappa}{v + \kappa} \right)^{\kappa} \left[1 - \left(\frac{v}{v + \kappa} \right) z \right]^{-\kappa}$$

giving

$$G_X(\mathbf{z}) = E[G_X(\mathbf{z}|N)] = \left(\frac{\kappa}{\kappa + v} \right)^{\kappa} \left[1 - \left(\frac{v}{v + \kappa} \right) (\Sigma p_i z_i) \right]^{-\kappa}.$$

This is the p.g.f. for the so-called multivariate negative binomial distribution (Patil and Joshi, 1968, p. 71). The marginal p.g.f. for X_i is

$$G_{X_i}(z) = \left(\frac{\kappa}{\kappa + p_i v}\right)^{\kappa} \left[1 - \left(\frac{p_i v}{p_i v + \kappa}\right)z\right]^{-\kappa},$$

showing that X_i is negative binomially distributed with mean vp_i and shape parameter κ. The correlation between X_i and $X_j, i \neq j$, is

$$\rho(X_i, X_j) = \left[\left(\frac{\kappa}{vp_i} + 1\right)\left(\frac{\kappa}{vp_i} + 1\right)\right]^{-1/2}.$$

Example 2.6. Suppose that v is lognormally distributed with $E(\ln v) = \xi$, $\text{var}(\ln v) = \sigma^2$. Then N is Poisson lognormal with parameters (ξ, σ^2),

$$P_N(r) = \frac{(2\pi\sigma^2)^{-1/2}}{r!} \int_0^{\infty} v^{r-1} e^{-v} e^{-(\ln v - \xi)^2/2\sigma^2} dv.$$

Now, the distribution of X_i is also Poisson lognormal, but with parameters $(\xi + \ln p_i, \sigma^2)$. The correlation matrix is

$$\rho(X_i, X_j) = \prod_{t=i,j} \{1 + [p_t e^{\xi + (1/2)\sigma^2}(e^{\sigma^2} - 1)]^{-1}\}^{-1}$$

For the general mixture on v of the Poisson distribution with parameter v, where v has the m.g.f. $M_v(t)$, the p.g.f. for N is given by

$$G_N(z) = M_v(z - 1).$$

Hence

$$G_X(\mathbf{z}) = M_v(\Sigma p_i z_i - 1), \qquad G_{X_i}(z) = M_v[p_i(z - 1)].$$

Conditioning upon N or v, these models are equivalent for any choice of $M_v(t)$, Examples 2.5. and 2.6. representing special cases.

2.4.2. *Estimation*

It was shown in Section 2.3. that $q = \max_i(x_i)$ is the critical size of the sample to ensure the existence of an unbiased estimator of functions of the type

$$B = \sum_{i=1}^{s} b(x_i, n) = \sum_{j=1}^{n} r_j b(j, n).$$

As the value of n increases, q is also likely to be quite large. In view of this, it is unrealistic to think that $N \geqslant q$. In fact, since the expectation of any statistic is a polynomial of degree N in $p_1, p_2, \ldots, p_s$ for the multinomial model, these polynomials are the only functions of $\mathbf{p}$ that may possibly be estimated without bias. For example, Simpson's index of diversity is $H_s = 1 - \Sigma p_i^2$, and the unbiased estimator is

$$\hat{H}_s = 1 - \sum \frac{X_i(X_i - 1)}{N(N - 1)} = 1 - \sum \frac{R_i i(i - 1)}{N(N - 1)}$$

which is the same as for the finite population. In fact, $\hat{h}_s$ is also sufficient, and therefore the minimum variance unbiased estimator for h_s.

When functions other than polynomials are to be estimated, some difficulties arise, as illustrated by the following theorem.

Theorem 2.2. Let $\mathbf{z}$ be a random variable that can only take a finite number of values, and write $p(\mathbf{z}; \Theta)$, $\Theta \in \Omega$, for the distribution of $\mathbf{z}$. Let $\Psi(\Theta)$ be some function of Θ that is not uniformly bounded in Ω. Then, for any estimator $\hat{\Psi}$ for Ψ and any $M > 0$, there exists a population $\Theta^* \in \Omega$ so that $|E[\hat{\Psi} - \Psi(\Theta^*)]| > M$, where the expectation is taken over the population Θ^*.

Proof. Let $\hat{\Psi}_1, \hat{\Psi}_2, \ldots, \hat{\Psi}_m$ be the possible values of $\hat{\Psi}$. If $\max_i |\hat{\Psi}_i| = \infty$ then $|E(\hat{\Psi} - \Psi)| > M$ in the case that $\Psi(\Theta^*)$ is finite. Otherwise $|E(\hat{\Psi} - \Psi)| > M$ by chosing Θ^* so that $\Psi(\Theta^*) > M + \max_i |\hat{\Psi}_i|$.

The following are examples of functions that cannot be estimated with controlled bias when $\mathbf{X}$ is a multinomial variate and $\mathbf{R}$ is the only observable statistic: s (the number of classes); $- \Sigma p_i \ln p_i$ (Shannon's information index, Shannon and Weaver, 1948); $(\Sigma p_i^2)^{-1}$; $(1 - \Sigma p_i^2)^{-1}$; $\Sigma p_i^{-\beta}$ where $\beta > 0$; $[\mathrm{Inf}(p_i)]^{-1}$; and $[1 - \sup(p_i)]^{-1}$.[2] Since a small standard error cannot correct a large bias, it is rather useless to search for efficient estimators of unbounded functions of $\mathbf{p}$. Some method of keeping the bias under control is called for and this cannot be done without imposing further constraints on the model. This is one theoretical reason for introducing abundance models. This will be done in Chapter 3.

2.5. Quadrat sampling: presence absence data

In some practical situations it may be impossible to count the number of elements of the various classes appearing in the sample.

In botanical ecology, for example, one is often faced with the problem of defining or determining what is an individual plant. When this is impossible (or difficult), sampling is carried out by placing a number of quadrats, or some other samples of fixed area and shape, more or less randomly in the field to be investigated. Since the elements (individuals) cannot be counted, only the presence or absence of the various classes (species) is recorded for each quadrat. The term 'qualitative data' is sometimes used for this sampling scheme, while 'quantitative data' refers to the complete counts of the number of elements.

Now let Y_i denote the number of quadrats where C_i is present. If there are Q quadrats placed at random, then

$$Pr(Y_i = r) = \binom{Q}{r} [1 - u_0(\lambda_i, \theta)]^r u_o(\lambda_i, \theta)^{Q-r},$$

where $u_j(\lambda_i, \theta)$ is the probability that C_i is represented by j elements in one particular quadrat. Let M_r be the number of classes represented in exactly r quadrats. Then

$$E(M_r|\lambda) = \sum_{i=1}^{s} \binom{Q}{r} [1 - u_0(\lambda_i, \theta)]^r u_0(\lambda_i, \theta)^{Q-r},$$

which, adopting the notation of Section 2.2., can be expressed in the form

$$E(M_r) = s\binom{Q}{r} E\{[1 - u_0(\Lambda, \theta)]^r u_0(\Lambda, \theta)^{Q-r}\} \qquad (2.13)$$

or from Example 1.11.

$$E(M_r) = \binom{Q}{r} \mu E\left\{ \frac{[1 - u_0(\Lambda^*, \theta^*)]^r \mu_0(\Lambda^*, \theta^*)^{Q-r}}{\Lambda^*} \right\} \qquad (2.14)$$

Further theoretical treatment and applications will be dealt with in Chapter 7.

2.6. Species–area curves

We shall, mainly for historical reasons, consider what in biology are known as species–area curves. The very first attempts to analyse populations of many classes were in fact attempts to plot the number of species found in an area as a function of the size of the area. Since the expected number of individuals in an area is proportional to the size of the area, it makes little difference whether we consider

$E(S)$ as a function of the area or as a function of N or v. For the multinomial model

$$E(S) = \sum_{i=1}^{s} [1 - (1 - p_i)^N],$$

or in terms of P^*,

$$E(S|N) = E\left\{\frac{[1 - (1 - P^*)^N]}{P^*}\right\}. \tag{2.15}$$

Example 2.7. For the models considered in Example 2.3. the structural distribution is $h^*(p) = \alpha(1 - p)^{\alpha - 1}$ Then, from (2.15)

$$E(S|N) = \int_0^1 \left[\frac{1 - (1 - p)^N}{p}\right]\alpha(1 - p)^{\alpha - 1}dp = \alpha[\psi(\alpha + N) - \psi(\alpha)]$$

where $\psi(\cdot)$ denotes the digamma function.

If we apply the expansion $\psi(z) = \ln z - \dfrac{1}{2z} + O(z^{-2})$, the relation may be written

$$E(S|N) = \alpha \ln\left(\frac{\alpha + N}{\alpha}\right) - \frac{\alpha}{2(\alpha + N)} + \frac{1}{2} + O(\alpha^{-1}),$$

showing that $E(S|N)$ is asymptotically proportional to $\ln N$.

Example 2.8. For MacArthurs's broken stick model the λ_i are supposed to be independently exponentially distributed with the same parameter. Then **p** has the Dirichlet distribution given in Example 1.6. with $\kappa = 1$ and $h^*(p) = s(s - 1)p(1 - p)^{s - 2}$. From (2.15) it follows that

$$E(S|N) = \int s(s - 1)p(1 - p)^{s - 2}\left\{\frac{[1 - (1 - p)^N]}{p}\right\}dp = \frac{sN}{N + s - 1}.$$

The problem of estimating points on the species–area curve has been considered by Krylow (1970), Holthe (1975) and Engen (1976a). Unbiased estimators of $E(S|N)$, for N larger than the actual value observed, do not exist. This is a fact strengthening the necessity of imposing further constraints on the model.

2.7. Good's (empirical) bayesian approach

Before considering special hypotheses related to the vector of abundances, we shall review the main results presented in an interesting paper by Good (1953). His work is regarded as one of the precursors of what is known as 'empirical Bayes methods' (Maritz, 1970).

Suppose that $n = \infty$, while s is finite. According to Good, the results would be virtually unchanged if s were enumerably infinite, but the proofs are more rigorous when it is finite. Our aim is to estimate the components of $\mathbf{p}$. Suppose that in a sample of size N one particular class occurs j times ($j = 0, 1, 2, ...$). We shall consider the final (posterior) probability that this class is the class with abundance $p_i, i = 1, 2, ..., s$. For the sake of rigour it is necessary to define precisely how the classes are selected for consideration. We shall suppose that they are sampled equiprobably from the s classes in such a way that the initial (prior) probability of chosing the i'th species is $1/s$. Further, let q denote the relative abundance of this class and let A_j be the event that it is represented by exactly j elements in the sample. Then, using Bayes' theorem, we have

$$Pr(q = p_i | A_j) = \frac{Pr(A_j | q = p_i)Pr(q = p_i)}{\sum\limits_{k=1}^{s} P(A_j | q = p_k)Pr(q = p_k)}, \qquad i = 1, 2, ..., s.$$

$$= \frac{p_i^j(1 - p_i)^{N-j}}{\sum\limits_{k=1}^{s} p_k^j(1 - p_k)^{N-j}} \qquad (2.16.)$$

Hence the moments of q come out as

$$E(q^m | A_j) = \frac{\sum\limits_{k=1}^{s} p_k^{j+m}(1 - p_k)^{N-j}}{\sum\limits_{k=1}^{s} p_k^j(1 - p_k)^{N-j}} \qquad (2.17)$$

We shall write $E_N(R_j)$ for the expectation of R_j in a sample of size N from $\mathbf{p}$. Hence

$$E_N(R_j) = \sum\limits_{k=1}^{s} \binom{N}{j} p_k^j(1 - p_k)^{N-j} \qquad (2.18)$$

and (2.18) may be written

$$E(q^m | A_j) = \frac{(j + m)^{(m)} E_{N+m}(R_{j+m})}{(N + m)^{(m)} E_N(R_j)} \qquad (2.19)$$

where $t^{(m)} = t(t-1)\ldots(t-m+1)$, and in particular

$$E(q \mid A_j) = \frac{j+1}{N+1} \frac{E_{N+1}(R_{j+1})}{E_N(R_j)} \qquad (2.20)$$

which is Good's basic result. The expected total abundance of all classes appearing j times in the sample is

$$E\left[R_j \frac{j+1}{N+1} \cdot \frac{E_{N+1}(R_{j+1})}{E_N(R_j)} \right] = \frac{j+1}{N+1} E_{N+1}(R_{j+1}). \qquad (2.21)$$

Now assuming that N is large, the total abundance of all classes appearing in the sample is approximately

$$\frac{1}{N} \sum_{j=1}^{\infty} E_N(R_{j+1})(j+1) = 1 - \frac{E_N(R_1)}{N}, \qquad (2.22)$$

and consequently (2.21) is valid also for $j = 0$, and R_1/N may serve as an estimator for the total abundance of those classes that do not appear in the sample.

Good discussed various methods for smoothing the sequence $R_1, R_2, \ldots$. Write $R'_1, R'_2, \ldots$ for such a 'smooth sequence'. Then $E(q \mid A_j)$ may be estimated by $\dfrac{j+1}{N+1} \cdot \dfrac{R'_{j+1}}{R'_j}$.

As a further illustration of the applicability of this theory consider now the population where all abundances are equal: $p_i = 1/s, i = 1, 2, \ldots, s$. If both s and N are large numbers of the same order, then

$$E_N(R_j) \approx s \frac{\beta^j}{j!} e^{-\beta},$$

where $\beta = N/S$.

In this case the smoothed sequence $R'_1, R'_2, \ldots$, will approximate the above Poisson sequence; hence, inserting R'_j and R'_{j+1} into (2.20), the estimate of q for a species occurring j times in the sample turns out to be a constant not depending on j, in accordance with the hypothesis. By using the standard procedure for the multinomial model, one would end up with a set of estimates that would differ in value, despite the fact that their expectations are equal.

For further applications see Good (1953), [See note].

Note:

The part of Good's approach that may be somewhat controversial is his assumption that the prior distribution is uniform. This distribution is related to some random experiment that is not likely to be carried out, since the species selected for consideration will in practice be those appearing in the sample. What seems to me to be a more realistic prior distribution is therefore

$$Pr(q = p_i) = C[1 - (1 - p_i)^N], \qquad i = 1, 2, \ldots, s,$$

where

$$C = \left\{ \sum_{i=1}^{s} [1 - (1 - p_i)^N] \right\}^{-1}.$$

Going through the same arguments with this prior distribution, equation (2.20) would be replaced by one involving terms like $E_{2N}(R_j)$. These terms are not easily estimated from the sample, a fact reducing the applicability of the theory. However, Good's application of Bayes' postulate is far more acceptable than is often the case in Bayesian theory. Note that the prior distribution

$$Pr(q = p_i) = 1/s, \qquad i = 1, 2, \ldots, s,$$

contains $(s - 1)$ unknown parameters which are all estimated from the sample.

CHAPTER THREE

Abundance models

3.1. General introduction

Theorem 2.2 demonstrated the necessity of introducing further constraints on the abundance model, that is, on the vectors λ and $\mathbf{p}$, in order to cope with the problem of estimating certain population parameters. The constraints are themselves properties of the population that it may be worthwhile to study for their own sake, not from the point of view of estimation but rather in descriptive or evolutionary terms. The present chapter deals with various constraints of deterministic (fixed) and stochastic nature, while the discussion of the realism of the various models in particular applications will be postponed to Part II of this monograph.

Example 2.5. (Continued). Consider again the heights of the adult male population in a city, with population mean $\bar{y}$ and variance σ_y^2. If Y is the height of one man chosen randomly from the population, a further description of the population would be included in the information that Y is approximately normally distributed. Then the number of people with a height less that y_0 is approximately $n_0 \Phi(y_0 - \bar{y})/\sigma_y$, where n_0 is the size of the population. Knowing $\bar{y}$ and σ_y, we may now construct a population which is approximately the same as the original one. Let the sequence $\xi_0, \xi_1, \ldots \xi_{n_0}$ be defined by

$$\Phi\left(\frac{\xi_i - \bar{y}}{\sigma_y}\right) = \frac{i}{n_0}$$

and put

$$y_i^* = n_0 \int_{\xi_{i-1}}^{\xi_i} x \frac{1}{\sqrt{(2\pi)}\sigma_y} e^{-(1/2)[(x-y)/\sigma_y]^2} dx, i = 1, 2, \ldots n_0.$$

Then $\bar{y}^* = \bar{y}$ exactly and the number of y_i^*s less than y_0 is

$$n_0 \Phi\left(\frac{y_0 - \bar{y}}{\sigma_y}\right) + \varepsilon, \text{ where } |\varepsilon| < 1.$$

Hence, the reconstructed population $y_1^*, y_2^*, \ldots, y_{n_0}$ is approximately the same as the original one, although it is given by the three parameters $\bar{y}, \sigma_y$ and n_0 only. The above constructed population is fixed; that is, the heights are not random variables. It is, however, more natural in this example to assume that heights are selected (by birth and growth of individuals) from an infinite population, so that the y_i's are independently normally distributed. Analogues to both of these constraints on the vector $(y_1, y_2, \ldots, y_{n_0})$ appear in the analysis of diversity. However, in the above example the distinction is not essential to the inference problem (at least when the sample size is small compared with n_0), while *it turns out to be quite crucial in the analysis of diversity*.

Examples of abundance models have already been given in Examples 2.3 and 2.4, where the distribution $h^*(p)$ was specified. Several other ways of defining models have been proposed in ecological literature, some of which will be reviewed in Part II. We shall concentrate on specifications of $h^*(p)$ or $g^*(\lambda)$ since this approach seems to cover most of the other proposals. The main purpose will be to estimate the parameters in $h^*(p)$ or $g^*(\lambda)$ from the observation $\mathbf{R} = (R_1, R_2, \ldots)$; further, it may be of interest to test the hypothesis that two populations possess the same structure or have some identical structural features that may be relevant to consider. Since we concentrate on $h^*(p)$ and not the simultaneous distribution of $(p_1, p_2, \ldots, p_s)$, it is important to bear in mind that quite different distributions of $\mathbf{p}$ may give identical or approximately identical structural distributions $h^*(p)$, as demonstrated in Examples 2.3 and 2.4. In the more familiar example with heights of people (example 2.5), the fact that the height of one person, chosen randomly from the population, was approximately normal gave rise to two different interpretations. The analogous distinction between two populations with approximately identical $h^*(p)$ will be considered next.

Definition 3.1. An abundance model is said to be symmetric if the distribution of $\mathbf{p} = (p_1, p_2, \ldots, p_s)$ is symmetric in all p_i (interchanging p_i and $p, i \neq j$, does not change the distribution).

For a symmetric model the distribution of P must be the same as the marginal distribution of any component p_i; that is, the distribution of p_i is $h(p)$. By Theorem 3.1 $h(p)$ does not exist when $s = \infty$; hence, symmetric models with an infinite number of classes do not exist.

Definition 3.2. We shall consider the construction of a fixed population by a known $h^*(p)$ which is continuous. This is analogous to the reconstruction of the approximately normal population in Example 2.5. The fixed population associated with $h^*(p)$ is given by the following construction. Write

$$s' = \int \frac{1}{p} h^*(p) dp$$

and put

$$s = \begin{cases} \text{smallest integer} \leqslant s' \text{ if } s' < \infty \\ \infty \text{ if } s' = \infty \end{cases}$$

Let the sequence $\xi_0, \xi_1, \ldots, \xi_s$ be defined by $\xi_0 = 1$,

$$\int_{\xi_i}^{\xi_{i-1}} \frac{1}{p} h^*(p) = 1, (i = 1, 2, \ldots, s - 1), \xi_0 = 0.$$

The fixed population is then given by

$$p_i = \int_{\xi_i}^{\xi_{i-1}} h^*(p) dp, i = 1, 2, \ldots, s.$$

Then

$$\sum_{i=1}^{s} p_i = 1$$

and the structural distribution of $\mathbf{p}$ is approximately $h^*(p)$. The model constructed in this way will be called 'the fixed model associated with $h^*(p)$.' Note that $h^*(p)$ need not necessarily range from 0 to 1, but may be positive also for $p > 1$. We may for example construct the fixed population associated with

$$h^*(p) = \alpha e^{-\alpha p}, p > 0.$$

If $u_j(vp_i, \theta)$ is the distribution of the number of elements sampled from C_i, then for the constructed fixed population

$$E(R_j) = \sum_{i=1}^{s} u_j(vp_i, \theta) \approx \int_0^1 u_j(vp, \theta) p^{-1} h^*(p) dp = E\left[\frac{u_j(vP^*, \theta)}{P^*}\right],$$

which is the same as (2.8), with the modification that the expectation is taken with respect to $h^*(p)$ instead of the exact structural distribution. The symmetric model and the fixed population associated with $h^*(p)$ represent two extremes from all populations with approximately the same $h^*(p)$. They are identical only if $p_1 = p_2 = \ldots = p_s = 1/s$. Examples of populations 'in between', that is, with random abundances with different distributions, have already been given (Examples 2.3 and 2.4).

3.2. Maximum likelihood theory applied to symmetric models

Suppose that the absolute abundances $\lambda_1, \lambda_2, \ldots, \lambda_s$ are independent- ' identically distributed with density $g(\lambda)$. This is clearly a symmetric model. Let $u_j(\lambda_i, \theta)$ be the distribution of the number of elements of C_i that appear in the sample, and let the classes be sampled independently. Then

$$P_{X_i}(x) = q_x(\xi) = \int_0^\infty u_x(\lambda, \theta) g(\lambda) d\lambda, \qquad i = 1, 2, \ldots, s,$$

where ξ is the vector of all parameters appearing in u_x and $g(\lambda)$. Without loss of generality assume that $X_1, X_2, \ldots, X_s > 0$, and $X_i = 0$ for $i > S$.
Then the likelihood is

$$L(X_1, X_2, \ldots, X_S | S) = \prod_{i=1}^{S} \left[\frac{q_{X_i}(\xi)}{1 - q_0(\xi)} \right] = \prod_{i=1}^{\infty} \left[\frac{q_i(\xi)}{1 - q_0(\xi)} \right]^{R_i}.$$

Consequently $(R_1, R_2, \ldots)$ is sufficient for ξ under the symmetric model if S is considered fixed (arguing conditionally upon S). We may now use classical maximum likelihood theory in this case, provided that we exercise a certain caution when interpreting the limit theorems (Anscombe, 1951). Note that S corresponds to the number of observations in classical maximum likelihood theory and we can only in theory imagine $S \to \infty$. In a particular applications, S can never exceed s, which is supposed to be finite. Another essential problem is that S cannot increase without increasing the sample size v. But then $g(\lambda)$ changes as well, and so does ξ. Parameters changing value with the number of observations are definitely outside the scope of classical maximum likelihood theory.

We shall also need the following limiting property of the distribution of $(R_1, R_2, \ldots)$ for the above model.

Let the expected sample size μ be fixed. Then $E(\Sigma X_i) = \mu$ and $sE(X_i) = \mu$, implying $\lim_{s \to \infty} q_0 = 1$ and $\lim_{s \to \infty} q_j = 0, j \geqslant 1$.

Further, since

$$E(X_i) = \frac{\mu}{s} = \sum_{i=1}^{\infty} i q_i$$

we see that $\lim (q_i s)$ must be finite for $i \geq 1$. Now assume

$$q_i s \to \beta_i, i = 1, 2, \dots.$$

Then

$$G_{(R_1, R_2, \dots)}(\mathbf{z}) = (q_0 + \sum_{i=1}^{\infty} z_i q_i)^s = \left[1 + \frac{1}{s} \sum_{i=1}^{\infty} q_i s(z_i - 1) \right]^s.$$

Hence

$$\lim_{s \to \infty} G(\mathbf{z}) = \prod_{i=1}^{\infty} e^{z_i(\beta_i - 1)}.$$

That is, in the limit, the R_i, are, for $i \geq 1$, independent Poisson variates with means β_i. A special case of this result was established by Anscombe (1950).

3.3. Fisher's logarithmic series model

In a study of Malayan butterflies, A.S. Corbet (1942) discovered that the sequence $R_1, R_2, \dots$ resembled the sequence $\alpha 1^{-m}, \alpha 2^{-m}, \alpha 3^{-m}, \dots$ By considering only the first 24 terms in a large sample he found a reasonably good fit by choosing $m = 1$. One year later, a collaboration with R.A. Fisher and C.B. Williams (Fisher, Corbet and Williams, 1943) resulted in a paper that deserves to be called a classic in statistical ecology. However, Fisher's statistical contribution, which will be dealt with below, has created much controversy and misunderstanding among research workers in statistics and ecology. This is due to the fact that Fisher did not put too much effort into explaining his assumptions and therefore opened up the possibility of different interpretations. In our terminology the model implies that $h^*(p) = \alpha(1 - p)^{\alpha - 1}$, as will be shown below. But it has already been shown in examples that quite different models of relative (or absolute) abundances can be associated with the same structural distribution $h^*(p)$, and this actually occurs where the controversy is hidden. Fisher's derivation goes as follows: suppose that the abundances $\lambda_1, \lambda_2, \dots, \lambda_s$ $(s < \infty)$ may be considered as s independent observations from the gamma distribution with parameters (ρ^{-1}, κ). For a given λ let $X_1, X_2, \dots, X_s$ be independent

Poisson variates with means $\lambda_1, \lambda_2, \ldots, \lambda_s$. Then the unconditional distribution of X_i is the negative binomial with parameters (ω, κ), where $\omega = (1 + \rho)^{-1}$. Since the zero class $(X_i = 0)$ is not observable*, Fisher considered the truncated form

$$\frac{\omega^k}{1 - \omega^k} \frac{\Gamma(\kappa + x)}{\Gamma(\kappa)x!} (1 - \omega)^x, x = 1, 2, \ldots.$$

Now, ρ must be proportional to the expected sample size, say $\rho = \mu\alpha^{-1}$, giving $\omega = \alpha/(\alpha + \mu)$, where α and κ now describe the distribution of the abundances in terms not depending on the actual sample size. The relative abundances possess the Dirichlet distribution

$$f(\mathbf{p}) = \frac{\Gamma(\kappa s)}{\Gamma(\kappa)^s} (p_1, p_2, \ldots, p_s)^{\kappa - 1}, \Sigma p_i = 1, \kappa > 0,$$

and, as shown in Example 2.6, the structural distribution for this model is the beta distribution with parameters $(\kappa + 1, \kappa s - \kappa)$. In Example 3.2 we showed that

$$E(R_j) = s \frac{\omega^\kappa}{1 - \omega^\kappa} \frac{\Gamma(\kappa + j)}{\Gamma(\kappa)\Gamma(j)} (1 - \omega)^j$$

under the above assumptions. When fitting this sequence of expectations to the observed sequence $R_1, R_2, \ldots$, Fisher found that κ usually took values close to zero, and he therefore proposed considering the limit $\kappa \to 0$. However, by Example 1.9 we have the relation $s = \alpha/\kappa$, expressing just the fact that the total expected abundance is

$$E\left(\sum_{i=1}^{s} \lambda_i \right) = s\rho\kappa = s\mu \frac{\kappa}{\alpha} = \mu.$$

In order to maintain this total abundance (mean density of individuals per sampling unit), it is necessary to require that $s \to \infty$ as $\kappa \to 0$, so that $s\kappa \to \alpha > 0$. Then

$$E(R_j) \to \alpha \frac{(1 - \omega)^j}{j}, j = 1, 2, \ldots \qquad (3.1)$$

and

$$h^*(p) \to \alpha(1 - p)^{\alpha - 1}. \qquad (3.2)$$

---

* This is in agreement with our general assumption in Chapter 2 that $\mathbf{X} = (X_1, X_2, \ldots, X_s)$ is not observable, but only $\mathbf{R} = (R_1, R_2, \ldots)$.

Since $E(R_j)$ is proportional to the terms of the logarithmic expansion, the model has become known as Fisher's logarithmic series model. It should be underlined that it is not legitimate to put $\kappa = 0$ and $s = \infty$ if the symmetric version of the model is considered, a fact following from Theorem 2.1. Asymmetric versions, however, such as that considered in Example 3.4, may be well defined also for $\kappa = 0$, $s = \infty$. The expected number of classes represented in the sample is

$$E(S) = \sum_{i=1}^{\infty} E(R_j) = \alpha \ln\left(\frac{\mu + \alpha}{\alpha}\right). \tag{3.3}$$

For large values of μ, ω is approximately zero and $E(R_j) \approx \alpha/j$ for j not too large. Hence, the model was in agreement with Corbets findings. Secondly, equation (3.3) 'explained' a relation proposed in botanical ecology by Gleason as early as 1922.

Parameter α, which is the only population parameter, was called the index of diversity. There is a considerable amount of data in the literature with which the logarithmic series fits well, which is surprising in view of the complexity of biological populations and the simplicity of this one-parameter model. It is the author's personal view, however, that the applicability of the model has been somewhat exaggerated, the reason being, perhaps, that research workers in ecology have found its simplicity alluring.

There is also a large number of statistical papers dealing with the logarithmic series model, only some of which can be mentioned here. Readers who want to study this literature in more detail should begin by consulting the papers by Boswell and Patil (1971) and Watterson (1974). In the field of ecology C.B. Williams wrote a series of papers on various applications of the model (Williams, 1944, 1946, 1947, 1950). He also published a monograph on this topic (Williams, 1964).

Maximum likelihood estimation

For the symmetric version we have

$$\lim_{k \to 0} Pr(X_i = x | X_i > 0) = \gamma \frac{(1 - \omega)^x}{x} = u_x(\omega), x = 1, 2, \dots.$$

where $\gamma = (-\ln\omega)^{-1}$, and the X_i are independent. If we therefore assume that κ is slightly larger than zero ($\kappa = 0$ is not allowed), then the positive X_i may be regarded as independent observations from the logarithmic series distribution, if S, the number of classes in the sample, is considered fixed. From Section 3.2 we know that **R** is sufficient for ω when the model is symmetric and further that the likelihood function is $\prod_{i=1}^{\infty} [u_i(\omega)]^{R_i}$. This would also be the likelihood function if our model was based on some random reordering of the abundances, such as that shown in Example 3.4. Then κ may be exactly zero, and the theory is no longer approximate. The maximum likelihood equation is

$$\sum_{i=1}^{\infty} R_i \frac{1}{u_i(\omega)} \frac{\partial u_i(\omega)}{\partial \omega} = 0 \tag{3.4}$$

giving

$$\hat{\omega} = \Phi^{-1}\left(\frac{S}{N}\right), \tag{3.5}$$

where

$$\Phi(x) = \frac{-x\ln x}{1 - x}, \qquad 0 < x < 1.$$

However, our primary interest is not in ω but in the population parameter $\alpha = \mu\omega/(1 - \omega)$, which is not dependent on the expected sample size μ. But conditionally upon S, **R** does not contain any information about α. Hence to find the maximum likelihood estimator for α, we must consider the unconditional model, following Anscombe (1950).

Several authors have dealt with the estimation of the parameter ω in the logarithmic series distribution. The maximum likelihood solution $\hat{\omega} = \Phi^{-1}$ (S/N) has been tabulated by Fisher (1943), Williamson and Bretherton (1964), and Patil and Wani (1965). The last authors also demonstrated that the bias of $\hat{\omega}$ is negative and they also found an expression for it. One of their expressions for the variance is

$$\text{var}\,\hat{\omega} = \omega^2(1 - \omega)/\{S\gamma[1 - \gamma(1 - \omega)]\} + O(S^{-2}),$$

recalling that $\gamma = -(\ln\omega)^{-1}$. They also found an expression for the term of order S^{-2}. The minimum variance unbiased estimator was discussed by Patil and Bildiker (1966). It takes the form

$$\hat{\omega}(N, S) = \begin{cases} 1 & \text{if } N = S \\ 1 - |S_{N-1}^{(S)}|/|S_N^{(S)}|, & \text{if } N > S, \end{cases}$$

where $S_N^{(S)}$ are the Stirling numbers of the first kind (see for example Abramowitz and Stegun (1965, Section 24.13). They gave tables for this estimator;

its variance may be estimated by

$$[1 - \hat{\hat{\omega}}(N,S)][\hat{\hat{\omega}}(N-1,S) - \hat{\hat{\omega}}(N,S)].$$

See also Bowman and Shenton (1970), Patil (1962) and Watterson (1974).

Let us now consider Anscombes derivation of the maximum likelihood estimator for α for the symmetric model (which is also valid for some reordering of this). From our limiting result at the end of Section 3.2 it follows that the R_i are independent Poisson variates, with $E(R_i) = \alpha(1 - \omega)^i/i$. Hence the log-likelihood function is

$$L = \alpha \ln \omega + S \ln \alpha + N \ln(1 - \omega) - \sum_{j=1}^{\infty} \{R_j \ln j + \ln R_j!\}.$$

Thus S and N are sufficient for estimating α and ω, and the maximum likelihood equations are

$$S = -\hat{\alpha} \ln \hat{\omega} \tag{3.6}$$

$$S = \hat{\alpha}[\ln(\hat{\alpha} + N) - \ln \hat{\alpha}]. \tag{3.7}$$

By inverting the matrix of expectations of the second derivatives of L to find the dispersal matrix of the estimates, we get

$$(-\ln \omega - 1 + \omega)\,\text{var}\,(\hat{\omega}) \approx \frac{1}{\alpha}(1 - \omega)\omega^2 \ln \omega,$$

$$(-\ln \omega - 1 + \omega)\text{cov}(\hat{\alpha}, \hat{\omega}) \approx -\omega(1 - \omega),$$

$$(-\ln \omega - 1 + \omega)\text{var}\,(\hat{\alpha}) \approx \alpha.$$

We are interested in an asymptotic formula for var $\hat{\alpha}$ as the sample size μ approaches infinity, while α is kept constant. Formally, from the last equation we have

$$\text{var}\,(\hat{\alpha}) = \frac{\alpha}{-\ln \omega + o(1)},$$

but as noted Section 3.2, we are not in the ordinary maximum likelihood situation and must exercise caution when interpreting limiting theory. However, Anscombe (1950) examined the simultaneous distribution of S and N and showed that the solution of equation (3.7) is asymptotically normally distributed and that the above formula for var $\hat{\alpha}$ is correct to the order suggested.

It is worth noticing that the variance of $\hat{\alpha}$ is of order $(\ln \mu)^{-1}$ in the sample size μ. Thus by increasing μ from, say, 1000 to 10 000, the standard error of $\hat{\alpha}$ is only reduced by 13%. In view of this, it is a waste of time to exaggerate the sample size if the sole purpose is to estimate α, provided that the model under discussion is realistic in the actual situation.

The fixed version

Consider now the fixed population associated with $h^*(p) = \alpha e^{-\alpha p}$, and let $\mathbf{p} = (p_1, p_2, \dots)$ be the relative abundances of this population. The construction of $\mathbf{p}$ is as in definition 3.2, and

$$E(R_j) \approx \alpha \frac{(1 - \omega)^j}{j}, j = 1, 2, \dots.$$

To find the distribution of $(R_1, R_2, \dots)$, we would have to insert $p_1, p_2, \dots$ in equation (2.3). This procedure would undoubtedly result in an utterly intractable expression which could not form the basis for the construction of estimators. Fisher (1943) proposed estimating α by equation (3.7) and evaluated an approximation for the variance of the estimator for the fixed model. This he called 'the variance of $\hat{\alpha}$ in parallel samples', an expression indicating that the abundances $\lambda_1, \lambda_2, \dots$, were kept constant; in other words, the only random element taken into account was that due to the random sampling of individuals. The approximation to var $\hat{\alpha}$ was found by the standard method, that is, by inserting $S = E(S) + \delta S$, $N = E(N) + \delta N$, $\hat{\alpha} = \alpha + \delta \hat{\alpha}$ and expanding to the first order in δS, δN and $\delta \hat{\alpha}$. Note that $\hat{\alpha}$ is expanded about the true value α, not about $E(\hat{\alpha})$. Then

$$E(\delta \hat{\alpha})^2 = \text{var}(\hat{\alpha}) + (\alpha - E\alpha)^2.$$

We shall use the notation var* $(\hat{\alpha})$ for this total mean square error. To find numerical values for var* $(\hat{\alpha})$, we need approximation to var (S) and cov (N, S), which are obtainable as follows. Write

$$S_i = \begin{cases} 1 & \text{if } X_i > 0 \\ 0 & \text{if } X_i = 0. \end{cases}$$

Then the S_i are independent and $\Sigma S_i = S$.
Now

$$\text{var}(S_i) = e^{-\mu p_i}(1 - e^{-\mu p_i}),$$

giving from Example 1.9.

$$\text{var}(S) = \sum_{i=1}^{\infty} (e^{-\mu p_i} - e^{-2\mu p_i}) = \int (e^{-\mu p} - e^{-2\mu p}) \frac{1}{p} h^*(p|\mathbf{p}) dp,$$

where $h^*(p|\mathbf{p})$ is the exact (discrete) structural distribution. Approximating $h^*(p|\mathbf{p})$ by $h^*(p)$, the integral is easily solved, giving

$$\text{var}(S) \approx \alpha \ln(2 - \omega).$$

Similar calculations yield

$$\text{cov}(N, S) \approx \mu\omega,$$

and bearing in mind that N is a Poisson variate, $\text{var}\, N = \mu$. Inserting these results in the squared equation after taking expectations, we get

$$\text{var}^*(\hat{\alpha}) \approx \frac{\mu\omega \ln(2 - \omega)}{\{1 - \omega + \ln \omega\}^2 (1 - \omega)}.$$

Fisher also proposed conditioning upon N, which to the first order (normal approximation) gives

$$\frac{1}{N} \text{var}^*(\hat{\alpha}|N) \approx \frac{\{1 - \omega(1 - \omega)/\ln(2 - \omega)\} \omega \ln(2 - \omega)}{\{1 - \omega + \ln \omega\}^2 (1 - \omega)}.$$

For large sample sizes ω vanishes compared with 1. Hence we are left with the simple formulae

$$\text{var}^*(\hat{\alpha}) \approx \text{var}^*(\hat{\alpha}|N) \approx \frac{\alpha \ln 2}{(1 + \ln \omega)^2}.$$

This is of order $(\ln \mu)^{-2}$, while for the symmetric model the same estimation equation resulted in a variance of order $(\ln \mu)^{-1}$. However, there is no reason to believe that (3.7) should be a preferable estimation equation for the fixed model. Let us look at estimation methods symbolized $\{a_i, b_i\}$, with estimation equations

$$\Sigma a_i R_i = E^*(\Sigma a_i R_i), \ \Sigma b_i R_i = E^*(\Sigma b_i R_i)$$

where E^* indicates that the parameters in the expressions for the expectations are replaced by the corresponding estimators. It is a simple matter to show that Fisher's method is equivalent to $\{1, i\}$. The methods $\{(i + 1)^{-1}, 1\}, \{i, i^2\}$ and $\{(i + 1)^{-1}, i\}$ have been investigated by the author (Engen, 1974). Only $\{i, i^2\}$ turned out to give smaller variance than $\{1, i\}$ for values of ω that are relevant.

The estimator for $\{i, i^2\}$ takes the form

$$\hat{\alpha} = N^2/(\Sigma i^2 R_i - N),$$

and to the first order

$$\frac{1}{N}\, \text{var}^*(\hat{\alpha}) \approx \frac{2\omega^2(2-\omega)}{(1-\omega)^2}.$$

Hence, $\text{var}^*(\hat{\alpha})$ is of order μ^{-1} and is preferable for large values of μ. However, for $\omega > 0.018$ Fisher's estimate turns out to be more precise (to this order of approximation).

3.4. The negative binomial model

Suppose that $\lambda_1, \lambda_2, \ldots, \lambda_s$ where $s < \infty$, are independently and identically gamma distributed with parameters (ρ^{-1}, κ). If this is the case, ρ must be proportional to the sample size, say $\rho = \mu/\alpha$, and, since $E(\Sigma \lambda_i) = s\rho\kappa = \mu$, we have the relation $s\kappa = \alpha$. The parameters α and κ are now unaffected by changes in the sample size μ. On the basis of Poisson sampling, we showed in Example 2.2 that

$$E(R_j) = s\left(\frac{1}{\rho+1}\right)^\kappa \frac{\Gamma(\kappa+j)}{\Gamma(\kappa)j!}\left(\frac{\rho}{\rho+1}\right)^j = \alpha\omega^\kappa \frac{\Gamma(\kappa+j)(1-\omega)^j}{\Gamma(\kappa+1)j!},$$

$$j = 0, 1, 2 \ldots \tag{3.8}$$

where $\omega = \alpha/(\alpha+\mu)$. In Example 2.6, we showed that the structural distribution for this model is the beta distribution with parameters $(\kappa+1, \alpha-\kappa)$. We have seen in the previous section that the logarithmic series is obtained by letting $\kappa \to 0$ with α kept constant, implying that $s \to \infty$. However, from (3.8) and the structural distribution for this model, it appears that the natural lower bound for k seems to be -1 (Engen, 1974) and not zero as claimed by Fisher (1943). Take a sequence of symmetric populations $\{\mathbf{p}_i\}$. Let $E(R_j^{(i)})$ denote the value of $E(R_j)$ for the population $\mathbf{p}_i$. We shall now demonstrate the existence of sequences of symmetric populations with the property that

$$\lim_{i \to \infty} E(R_j^{(i)}) = \alpha\omega^\kappa \frac{\Gamma(\kappa+j)}{\Gamma(\kappa+1)j!}(1-\omega)^j, j = 1, 2, \ldots. \qquad \text{if } -1 < \kappa < 0.$$

There are several ways of constructing such sequences; the following is an example that takes a rather tractable mathematical form.

Let $\mathbf{p}_i$ be the symmetric model where the λ_i possess the distribution

$$\frac{\alpha^\kappa}{\Gamma(\kappa)}\frac{1}{(1-\varepsilon^\kappa)}(1-e^{-i\lambda})\lambda^{\kappa-1}e^{-\alpha\lambda}, \qquad i=1,2,\dots$$

where $\varepsilon_i = \alpha/(\alpha+i)$ and $\kappa > -1$.

Compounding this with the Poisson distribution in the usual way, and scaling so that the total abundance is kept constant, we get

$$E(R_j^{(i)}) = \frac{\alpha\omega^\kappa\Gamma(\kappa+j)(1-\omega)^j(1-\eta_i^{\kappa+j})}{(1-\varepsilon_i^{\kappa+1})\Gamma(\kappa+j)j!}, \qquad j=0,1,2\dots$$

where $\eta_i = (\alpha+\mu)/(\alpha+\mu+i)$. Hence

$$\lim_{i\to\infty} E(R_j^{(i)}) = \begin{cases} \infty & \text{for } j=0 \\ \dfrac{\alpha\omega^\kappa\Gamma(\kappa+j)}{\Gamma(\kappa+1)j!}(1-\omega)^j & \text{for } j=1,2,\dots. \end{cases}$$

which is the negative binomial series with $\kappa > -1$. Also, for any class,

$$Pr(X_j = x \mid X_j > 0) \to \frac{\omega^\kappa}{1-\omega^\kappa}\frac{\Gamma(\kappa+x)(1-\omega)^x}{\Gamma(\kappa+1)x!}, \qquad x=1,2,\dots.$$

For sufficiently large values of i, the positive X_j constitutes a set of approximately truncated negative binomially distributed random variables with κ possibly in the interval $\langle -1, 0]$.

Note that, according to Theorem 2.1, symmetric models with an infinite number of classes do not exist. Hence it is not legitimate to put $i=\infty$ and $s=\infty$.

From our limiting result at the end of Section 3.2, it follows that $(R_1, R_2, \dots)$ tends to be independently Poisson distributed as $i \to \infty$. It is easy to check that

$$\lim_{i\to\infty} g^*(\lambda) = \frac{\alpha^{\kappa+1}}{\Gamma(\kappa+1)}\lambda^\kappa e^{-\alpha\lambda}.$$

A reordering model

For positive values of κ (and $s < \infty$), the simultaneous density of $p_1, p_2, \dots, p_s$ for the symmetric model is the Dirichlet density

$$f_\mathbf{p}(\mathbf{p}) = \frac{\Gamma(\kappa s)}{\Gamma(\kappa)^s}(p_1, p_2, \dots, p_s)^{\kappa-1}, \quad \Sigma p_i = 1.$$

In Example 2.4, we discussed a model with relative abundances

$(q_1, q_2, \ldots, q_s)$, which was some specified reordering of $(p_1, p_2, \ldots, p_s)$. The reordering was accomplished in such a way that

$$Q_i = q_i \bigg/ \left(1 - \sum_{j=1}^{i-1} q_j\right), \quad i = 1, 2, \ldots, s-1,$$

was beta distributed with parameters $(\kappa + 1, \alpha - \kappa i)$, and that the Q_is were independent. It was shown in Example 2.3 that this model is well defined also if $\kappa = 0$, in which case $\{E(q_i)\}_{i=1}^{\infty}$ forms a geometric sequence. Since the distribution of Q_i exists for any $\kappa > -1$, it seems worthwhile to make a close study of the model as a whole to see whether it is well defined when $\kappa < 0$, and if so, what its structural distribution is.

We first show that $\sum_{i=1}^{n} q_i$ tends to 1 in probability as $n \to \infty$. Since, by construction,

$$\sum_{i=1}^{n} q_i \leqslant 1,$$

it is sufficient to show that

$$\lim_{n \to \infty} E\left(\sum_{i=1}^{n} q_i\right) = 1.$$

Now

$$E(Q_i) = \frac{\kappa + 1}{\alpha - i\kappa + \kappa + 1} = \frac{a}{b + i - 1},$$

where $a = -(\kappa + 1)/\kappa, b = -(\alpha + 1)/\kappa$.
Hence

$$E(q_i) = E(Q_i) \prod_{j=1}^{i-1} [1 - E(Q_j)] = \frac{a\Gamma(b)\Gamma(b - a - 1 + i)}{\Gamma(b - a)\Gamma(b + i)}. \quad (3.9)$$

Now, $\sum_{i=1}^{\infty} E(q_i)$ turns out to be a special case of the sum of the terms of the negative binomial beta distribution defined in Section 1.1, and is therefore equal to 1. Hence, the q_i may denote relative abundances, and our model is well defined. Note that the sequence $\{E(q_i)\}_{i=1}^{\infty}$ tends to the geometric series as $\kappa \to 0_-$, just as in the case $\kappa \to 0_+$.

One will intuitively guess that the structural distribution is the beta distribution with parameters $(\kappa + 1, \alpha - \kappa)$ also when $\kappa \in < -1$,

$\kappa \varepsilon < -1, 0]$. Though this is correct, the proof which appears in Appendix A, is non-trivial.

The species–area curve.

According to (2.17) we have

$$E(S|N) = E\left\{\frac{1 - (1 - P^*)^N}{P^*}\right\}.$$

The case $\kappa = 0$ was dealt with in Example 2.8. Let us now assume that $\kappa \neq 0$. Evaluating the above expectation when P^* is a beta variable with parameters $(\kappa + 1, \alpha - \kappa)$, we get

$$E(S|N) = \frac{\alpha}{\kappa}\left[1 - \frac{\Gamma(\alpha)\Gamma(\alpha + N - \kappa)}{\Gamma(\alpha - \kappa)\Gamma(\alpha + N)}\right]. \tag{3.10}$$

In practice $|\kappa|$ will often be small compared with α and $\alpha + N$, in which case a simple Taylor expansion gives

$$E(S|N) \approx \frac{\alpha}{\kappa}\left\{1 - e^{-\kappa[\psi(\alpha - N) - \psi(\alpha)]}\right\} \tag{3.11}$$

where $\psi(\cdot)$ denotes the digamma function. Further, if α is not too small

$$E(S|N) \approx \frac{\alpha}{\kappa}\left\{1 - \left(\frac{\alpha}{\alpha + N}\right)^{\kappa}\right\}. \tag{3.12}$$

In any case the following asymptotic results hold:

$$\lim_{N \to \infty} E(S|N) = \frac{\alpha}{\kappa} = s \quad \text{if } \kappa > 0.$$

$$E(S|N) = \frac{\Gamma(\alpha + 1)}{-\kappa\Gamma(\alpha - \kappa)} N^{-\kappa} + 0(N^{-(\kappa + 1)}) \quad \text{if } \kappa < 0$$

(for the case $\kappa = 0$ see example 2.8).

Consequently, if $\kappa < 0$, then the logarithm of the number of classes is approximately proportional to the logarithm of the number of elements sampled. This relation has repeatedly been proposed by a number of authors dealing with ecology (Arrhenius, 1921; Kilburn, 1963, 1966).[†]

Estimation in the symmetric model

Consider now the symmetric model or some reordering of it. Then $[E(R_j)]$ is at least approximately proportional to the terms of the truncated negative binomial expansion. If $\kappa < 0$, then s is a large number, and from a general result in Section 2.2 the R_i are independent Poisson variates. For $\kappa > 0$, $\mathbf{R}$ possesses a multinomial distribution. In any case, arguing conditionally upon the observed number of classes, the M.L.E. of ω, κ is given by

$$N = \hat{\kappa}\Phi(\hat{\omega})^{-1} \sum_{i=1}^{\infty} R_i U_i(\hat{\kappa}) \tag{3.13}$$

$$M = N/S = \frac{\hat{\kappa}(1 - \hat{\omega})}{\hat{\omega}(1 - \hat{\omega}^{\hat{\kappa}})}, \tag{3.14}$$

where

$$\Phi(\omega) = \frac{-\omega \ln \omega}{1 - \omega}$$

and

$$U_i(\kappa) = \sum_{j=1}^{i} (\kappa + j - 1)^{-1}.$$

If we also estimate the sample size μ by the observed number of elements, that is, if we write $\hat{\mu} = N$, we have a set of equations giving estimators for the population parameters α, κ, writing $\hat{\alpha} = \hat{\mu}\hat{\omega}/(1 - \hat{\omega})$. The solution of (3.13) and (3.14) was discussed by Sampfort (1955) with reference to cases where $\kappa > 0$. The present author (Engen, 1974) examined the case where κ is allowed to take

---

[†] These curves have some historical interest. Arrhenius' proposal (Arrhenius, 1921) probably represents the first attempt to estimate parameters related to the concept of species diversity. However, his work was soon criticised by Gleason (1922), who proposed that S was approximately proportional to $\ln N$. Both equations were supported by field data. From the above discussion it appears that the two formulae merely represent different values of κ in the negative binomial model. Gleasons's formula corresponds to $\kappa = 0$, which is Fisher's model, while Arrhenius' formula is derived by assuming that $-1 < \kappa < 0$.

any value greater than -1. Write

$$\omega_1 = \Phi^{-1}(M^{-1})$$

and

$$\eta = \frac{1}{2s} \sum_{i=2}^{\infty} R_i U_{i-1}(1).$$

Sampford (1955) showed that $\eta + \ln \omega_1 < 0$ is a sufficient condition for the existence of a solution $\hat{\kappa} \in \langle 0, \infty], \hat{\omega} \in \langle 0, 1 \rangle$, and if $\eta + \ln \omega_1 < 0$, then there is a solution $\hat{\kappa} \in \langle -1, 0 \rangle, \hat{\omega} \in \langle 0, 1 \rangle$ (Engen, 1974). To find the solution in practice, it is convenient to start the iteration at $\kappa = 0$ in the direction given by the sign of $\eta + \ln \omega_1$.

Approximations to the standard errors may be found in the usual way by evaluating the components of the information matrix (see Sampford, 1955; Engen, 1975b). However, it is the properties of $\hat{\alpha}, \hat{\kappa}$ that are of interest from our point of view. In order to find these, we should need to generalize Anscombe's work on the logseries model (Anscombe, 1950) considered in Section 3.3. To the best of my knowledge this problem has not been dealt with in statistical literature.

The parameters κ and ω may also be estimated by considering the first two moments of the zero-truncated negative binomial distribution, though this is a less efficient method. Writing $T = \sum_{i=1}^{\infty} R_i i(i-1)$, the estimation equations are (3.14) combined with

$$T = \frac{(\hat{\hat{\kappa}} + 1)(1 - \hat{\hat{\omega}})}{\hat{\hat{\omega}}}. \tag{3.15}$$

Eliminating $\hat{\hat{\kappa}}$ we get

$$Me^{-T\Phi(\hat{\hat{\omega}})} = 1 - \hat{\hat{\omega}}(T - M + 1). \tag{3.16}$$

This equation has the false solution $\omega_0 = 1/(T + 1)$, which, inserted into (3.15) corresponds to $\kappa_0 = 0$. (These values do not fit into (3.14)). Sampford has shown that a solution $\hat{\hat{\kappa}} \in \langle 0, \infty], \hat{\hat{\omega}} \in \langle 0, 1 \rangle$ exists if $1 - e^{-T} < T/M < \ln(T + 1)$. Neither of the above inequalities is necessarily satisfied if $\kappa > 0$. For the extended model ($\kappa > -1$), however, it is a simple matter to check that $1 - e^{-T} < T/M$ is a sufficient condition for the existence of a unique solution, and that

$$T/M \lessgtr \ln(T + 1) \Rightarrow \hat{\hat{\omega}} \lessgtr \omega_0, \hat{\hat{\kappa}} \gtrless 0.$$

Again, it is convenient to start interations at $\kappa = 0$, this time in the direction given by the sign of $T/M - \ln(T+1)$.

The author has evaluated approximations for the efficiency of the moment method for values of k and ω likely to occur in practice (at least for ecological applications). Writing E for the efficiency, we have for example

$$\left.\begin{array}{l} \kappa = 2 => E < 0.68 \\ \kappa = 1 => E < 0.62 \\ \kappa = 0 => E < 0.51 \\ \kappa = -\tfrac{1}{2} => E < 0.44 \end{array}\right\} \quad \text{for all } \omega \in [0.001, 0.1].$$

Estimation in the fixed model

Let us now consider the fixed model associated with

$$h^*(p) = \frac{\alpha^{\kappa+1}}{\Gamma(\kappa+1)} p^\kappa e^{-\alpha p}, \quad \text{say } \mathbf{p}(\alpha, \kappa) = \{p_1, p_2, \ldots, p_s\},$$

giving

$$E(R_j) \approx \alpha \omega^\kappa \frac{\Gamma(\kappa+j)(1-\omega)^j}{\Gamma(\kappa+1)j!}, \quad j = 1, 2, \ldots \dagger$$

(see definition 3.2, p. 36).

The distribution of $\mathbf{R}$ is now given by equation (2.3), where the p_i are the above functions of α and κ. We concluded in Chapter 2 that this distribution was too intractable to form a basis for proposing estimation methods. In fact, this would be the case even if the p_is were simple functions of a small number of population parameters. Since this does not apply, the situation is even more difficult to handle. We are therefore forced to propose methods of estimation quite intuitively.

There are three parameters to be estimated, and we need three equations. Write in general $\{a_{1j}(\theta), a_{2j}(\theta), a_{3j}(\theta)\}$ for the method

$$\sum_{j=1}^{\infty} a_{ij}(\hat{\theta})R_j = \sum_{j=1}^{\infty} a_{ij}E(R_j)(\hat{\theta}),$$

where $\theta = (\alpha, \kappa, \mu)$ and $\hat{\theta} = (\hat{\alpha}, \hat{\kappa}, \hat{\mu})$.

---

† One may find better approximations to $E(R_j)$, but bearing in mind that fixed models of this kind must in any case be considered as fairly rough approximations to reality, it does not seem worthwhile to exaggerate on this point.

Then the maximum likelihood equation of the previous sections may be rendered $\{1, j, U_j(\kappa)\}$, while the moment method is $\{1, j, j^2\}$ or equivalently $\{1, j, j(j-1)\}$. The words 'maximum likelihood' and 'moment' bear no relevance to the fixed model, but it seems natural to list these methods as candidates for further investigation. Rao (1971) used the terms 'pseudo moment method' and 'pseudo maximum likelihood method' in a similar context.

Another method investigated by the author (Engen, 1974) is $[(j+1)^{-1}, 1, j]$. In addition to equation (3.14) and $\hat{\mu} = N$, this leads to

$$\hat{\kappa} = \frac{V - (1 - M^{-1})\hat{\omega}/(1 - \hat{\omega})}{V - M^{-1}},$$

where

$$V = \frac{1}{N} \sum_{i=1}^{\infty} (i+1)^{-1} R_i.$$

Eliminating κ we get

$$\hat{\omega}^{(1 - 2VM)/(1 - VM)} e^{-\Phi(\hat{\omega})(M-1)/(1-VM)} - \frac{\hat{\omega}[1 - VM(M+1)] + VM}{M(1 - VM)}$$

$$= 0 \tag{3.19}$$

(Engen, 1974).

Equation (3.19), too, has a false solution, $\omega = \omega_2 = M^{-1}$, which, inserted in (3.18), gives $\kappa = 1$. It was shown by the author that there exists a solution $(\hat{\omega}, \hat{\kappa})$ yielding

$$\hat{\omega} \lessgtr \omega_2, \hat{\kappa} \lessgtr 1 \text{ if } V \gtrless M^{-1} + \frac{\ln M}{(1 - M)^2}.$$

Evaluation of standard errors

It is rather tedious to establish even the first order approximation to the standard errors for the fixed model. Expanding the estimation equations to the first order, squaring and taking expectations, one ends up with a set of equations involving terms like $\text{cov}(R_i, S)$, $\text{cov}(R_i, N)$ and $\text{cov}(R_i, R_j)$. Further, these appear in sums like $\sum_{i=1}^{\infty} U_i(\kappa) \, \text{cov}(R_i, N)$. We give one example to illustrate how the computations go:

Write

$$R_i^{(j)} = \begin{cases} 1 & \text{if } X_j = i \\ 0 & \text{otherwise} \end{cases}.$$

Then

$$R_i = \sum_j R_i^{(j)} \quad \text{and} \quad N = \sum_j X_j.$$

If we assume Poisson sampling, the classes are sampled independently and we have

$$\text{cov}(R_i, N) = \sum_j \text{cov}(R_i^{(j)}, X_j).$$

This sum may be written

$$\sum_j \left[i \frac{(\mu p_j)^i}{i!} e^{-\mu p_j} - i \frac{(\mu p_j)^i}{i!} e^{-\mu p_j} \mu p_j \right] = \sum_j C(p_j).$$

Approximating this sum by $\int C(p) \frac{1}{p} h^*(p) dp$ we finally get

$$\text{cov}(R_i, N) \approx \frac{\alpha}{\kappa} \omega^\kappa (1 - \omega) \left[\Phi_{i-1}(\kappa, 1 - \omega)(\kappa + i - 1) \right.$$

$$\left. - \Phi_i(\kappa, 1 - \omega)(\kappa + i) \right].$$

where

$$\Phi_i(x, y) = \frac{x(x+1)\dots(x+i-1)}{i!} y^i, 0 < y < 1, i = 1, 2, \dots.$$

and

$$\Phi_0(x, y) = 1.$$

Then

$$\sum U_i(\kappa) \text{cov}(R_i, N) \approx \frac{\alpha}{\kappa} \omega^\kappa (1 - \omega) \left\{ \Phi_0 U_1(\kappa)\kappa + \right.$$

$$\left. + \sum_{i=1}^{\infty} \Phi_i(\kappa, 1 - \omega)(\kappa + i)[U_{i+1}(\kappa) - U_i(\kappa)] \right\}$$

$$= \frac{\alpha}{\kappa} \omega^\kappa (1 - \omega) \left[1 + \sum_{i=1}^{\infty} \Phi_i(\kappa, 1 - \omega) \right] = \frac{\alpha}{\kappa}(1 - \omega)$$

since

$$\sum_{i=0}^{\infty} \Phi_i(x,y) = y^{-x}.$$

Some of the algebraic results needed were derived by Rao (1971), others by Engen (1974, 1975b). Fig. 1 indicates which one of the above methods seems preferable for various values of κ and ω likely to occur in practice. It turns out that the pseudo-moment method gives the smallest generalized variance for most of the actual values of κ and ω. The pseudo-M.L.E. is most efficient only in a very small region, and the 'relative efficiency' of the pseudo-moment method has a minimum value as large as about 0.93 in this region. Our third method is preferable for very small values of κ and large values of ω (small sample sizes μ). It may be a preferable method in case the model is fitted to linguistic data. Data given by Yule (1944) and Zipf (1932) indicate that we should expect to find small values of κ and large values of ω if the negative binomial series is fitted. Standard errors for these methods are tabulated by Engen (1974, 1975b).

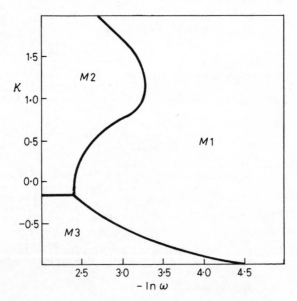

Fig. 1 *Diagram indicating the method with smallest generalized variance.* $M1 = \{1, j, j^2\}$. $M2 = \{(j+1)^{-1}, 1, j\}$. $M3 = \{1, j, U_j(\kappa)\}$

3.5. The geometric series model

In Example 2.3 we considered the model with relative abundances

$$q_1 = Z_1, q_i = Z_i \prod_{j=1}^{i-1} (1 - Z_j), \quad i = 2, 3, \dots,$$

where the Z_i were independent beta distributed variables with density $\alpha(1 - z)^{\alpha - 1}$. We showed that the structural distribution for this model was

$$h^*(p) = \alpha(1 - p)^{\alpha - 1},$$

which is the same as for Fisher's logarithmic series model. The expectations of the above abundances are

$$E(q_i) = \theta(1 - \theta)^{i-1}, i = 1, 2, \dots,$$

where $\theta = 1/(\alpha + 1)$. If we substitute the abundances by their expected values, we get the fixed model

$$p_i = \theta(1 - \theta)^{i-1}, i = 1, 2, \dots,$$

which, by the above argument, may be regarded as a fixed version of Fisher's model with $\alpha = 1/\theta - 1$. (Engen, 1975a).

The maximum likelihood estimate

Let the classes be recognizable and numbered $C_1, C_2, \dots,$ where the subscripts bear no relation to the ordering of the abundances (C_i is not in general the class with abundance p_i). If q_i is the abundance of C_i then

$$q_i = \theta(1 - \theta)^{u_i - 1}, \quad \text{where } \mathbf{u} = (u_1, u_2, \dots)$$

is the set of all positive integers in some order. According to our general assumptions stated in Chapter 2, $\mathbf{u}$ is unknown. As a sample of size N is drawn from the population, the log-likelihood is

$$L = \sum_{i=1}^{\infty} X_i \ln \left[\theta(1 - \theta)^{u_i - 1}\right].$$

Because the ordering of the p_is is the same for all values of $\theta, (p_1 > p_2 > \dots,) L$ is maximized for a given θ by putting u_i equal to the subscript of the ith most abundant class in the sample. Thus, if the order statistics of the X_i are $\mathbf{X}^* = [X^{(1)}, X^{(2)}, \dots]$ (note that $\mathbf{X}^*$ is observable, though $\mathbf{X}$ is not), and $\mathbf{T} = (t_1, t_2, \ldots)$ is defined by

$$X^{(1)} = X_{t_1} \geqslant X^{(2)} = X_{t_2} \geqslant \ldots,$$

the estimate of $\mathbf{u}$ is $\hat{\mathbf{u}} = \mathbf{T}$. Now, maximizing L with respect to $\alpha = (1 - \theta)/\theta$, we get the maximum likelihood estimate

$$\hat{\alpha}_N = \frac{1}{N} \sum_{i=1}^{\infty} iX^{(i)} - 1.$$

Let us now consider the bias of α_N. Define $Y = (Y_1, Y_2, \ldots)$ by

$$Y_{u_i} = X_i; \text{ then } E(Y_{u_i}) = E(X_i) = Np_{u_i}, \text{ or } E(Y_j) = Np_j.$$

If $\mathbf{u}$ were known, the maximum likelihood estimate of α would be $\tilde{\alpha}_N = \frac{1}{N} \sum_{i=1}^{\infty} iY_i - 1$, which is easily shown to be unbiased. Comparing $\tilde{\alpha}_N$ with $\hat{\alpha}_N$, we see that $\mathbf{Y}$ is replaced by $\mathbf{X}^*$ in the case that $\mathbf{u}$ is unknown. Intuitively, this may cause a considerable negative bias in $\hat{\alpha}_N$. Though the effect of correcting $\hat{\alpha}_N$ by Quenouille's jack-knife method (Quenouille, 1956) is unclear in this case,[†] one will intuitively tend to guess that jack-knifing is appropriate.

Write $\bar{\alpha}_{N-1}$ for the mean value of $\hat{\alpha}_{N-1}$ for all N subsamples of size $N - 1$. Then the corrected estimator is

$$\hat{\hat{\alpha}}_N = N\hat{\alpha}_N - (N - 1)\hat{\alpha}_{N-1}$$

Let $Z^{(1)} \geqslant Z^{(2)} \geqslant \ldots$ denote the ordered observed abundances in a random subsample of size $N - 1$. Then

$$\hat{\hat{\alpha}}_N = E\left(\sum_{i=1}^{\infty} i(X^{(i)} - Z^{(i)}) \right) - 1,$$

where E denote the mean value over all subsamples of size $N - 1$. Write

$$\Delta = \sum_{i=1}^{\infty} i(X^{(i)} - Z^{(i)}).$$

Then, if the element not included in the subsample belongs to a class originally represented by only one element in the sample, then $\Delta = S = \sum_{i=1}^{N} R_i$. This is due to the fact that all terms except the last (excluding those corresponding to $X^{(i)} = Z^{(i)} = 0$) would be zero.

---

† I owe thanks to C. Thaillie for pointing this out to me.

Hence

$$Pr\left(\Delta = \sum_{j=1}^{N} R_j\right) = R_1/N.$$

Proceeding in the same way, we see that

$$Pr(\Delta = S - R_1) = 2R_2/N,$$
$$Pr(\Delta = S - R_1 - R_2) = 3R_3/N,$$

and in general

$$Pr\left(\Delta = \sum_{j=1}^{N} R_j\right) = iR_i/N.$$

Consequently,

$$E(\Delta) = \frac{1}{N}\sum_{i=1}^{N} iR_i \sum_{j=i}^{N} R_j = \frac{1}{N}\sum_{i \leqslant j} iR_i R_j,$$

and

$$\hat{\hat{\alpha}}_N = \frac{1}{N}\sum_{i \leqslant j} iR_i R_j - 1.$$

3.6. The Poisson lognormal model

Introduction

Preston (1948) proposed grouping the observations in the following way ('logarithmic grouping'):

$$G_{-1} = R_0$$
$$G_0 = \tfrac{1}{2}R_1$$
$$G_1 = \tfrac{1}{2}R_1 + \tfrac{1}{2}R_2$$
$$G_3 = \tfrac{1}{2}R_2 + R_3 + \tfrac{1}{2}R_4$$
$$\vdots$$
$$G_j = \tfrac{1}{2}R_{2^{j-1}} + R_{2^{j-1}+1} + \ldots R_{2^j-1} + \tfrac{1}{2}R_{2^j}$$
$$\vdots$$

When plotting G_j against j for various ecological sets of data, Preston found 'curves' resembling the normal distribution truncated to the left. He argued that the underlying abundances were lognormally distributed and that the truncation was due to the fact that rare species were unlikely to appear in the sample. Grundy (1951) defined the model in statistical terms, assuming that the abundances $\lambda_i, i = 1, 2, \ldots, s$, were independently lognormally distributed and that the distribution of X_i conditionally upon λ_i was the Poisson distribution with mean λ_i. Accordingly $E(R_j)$ appeared to be proportional to the jth term of the Poisson lognormal distribution. Bulmer (1974) worked out the maximum likelihood estimators for this symmetric version of the model.

Now let the λ_i be independently lognormally distributed with parameter (ξ_0, σ^2) and assume that $se^{\xi_0 + (1/2)\sigma^2} = 1$, to ensure that ξ_0 and σ^2 are population parameters not depending on the sample size. Let the conditional distribution of X_i be the Poisson distribution with mean $\mu\lambda_i$, so that $E(\Sigma X_i) = E(N) = \mu$. Now X_i posesses the Poisson lognormal distribution with parameters $(\xi_0 + \ln \mu, \sigma^2)$ and $E(R_j) = sP_j(\xi, \sigma^2)$, where $P_j(\xi, \sigma^2)$ are the terms of the Poisson lognormal distribution with parameters $\xi = \xi_0 + \ln \mu$, and σ^2. It appears that the parameter ξ changes with the sample size, while σ^2 is independent of μ (Taylor and Kempton, 1974).

The fixed population associated with the lognormal distribution has not been dealt with in the literature.

The structural distribution

The distribution of the p_i is now rather intractable, and it is necessary to make some approximations. Omitting the index i, we have $\ln p = \ln \lambda - \ln \Sigma \lambda$. As $s \to \infty$, the distribution of $\Sigma \lambda$ tends to the normal by the central limit theorem, and so does the distribution of $\ln \Sigma \lambda$. $\ln p$ is therefore asymptotically normally distributed. In practice, however, s is not extremely large, while $v = \sigma^2$ is relatively large; hence the normal approximation is not necessarily valid. To obtain an improved solution let the cumulant generating function of $\ln p$ be approximated by

$$K(t) = k_1 t + k_2 t^2/2 + k_3 t^3/6 = \ln E(p^t),$$

where t is a generating symbol. The cumulants can be evaluated by solving

$$\ln E(p) = -\ln s = K(1),$$

$$\ln E(p^{-1}) = \ln s + v + \omega_1 = K(-1),$$

$$\ln E(p^{-2}) = 2\ln s + 3v + \omega_2 = K(-2),$$

where

$$\omega_1 = \ln\left[1 + (e^{-v} - 1)/s\right],$$

$$\omega_2 = \ln\left(1 + \left[(s-1)(e^v - 1) + 2(s-2)(e^{-2v} - 1) + (e^{-3v} - 1)\right]/s^2\right).$$

When s is relatively large we can use the approximations $\omega_1 = 0$, $\omega_2 = \ln\left[1 + (e^v - 1)/s\right]$. The solution of the above set of equations is

$$k_1 = -\ln s - v/2 - \omega_1 + \omega_2/6,$$

$$k_2 = v + \omega_1,$$

$$k_3 = 3\omega_1 - \omega_2,$$

which, inserted into the cumulant generating function, approximately specifies $g(p)$. Note, however, that the formula is only asymptotically correct either as $v \to 0$ or $s \to \infty$. We shall make use of this result in Chapter 5.

Maximul likelihood estimation in the symmetric model

From Bulmer (1974), the log-likelihood is

$$L = \sum_{j=1}^{\infty} R_j \ln\left\{P_j/(1 - P_0)\right\}$$

$$= \Sigma R_j \ln P_j - S \ln(1 - P_0).$$

Writing $\theta_1 = \xi, \theta_2 = \sigma^2$, the ML-score with respect to θ_i is

$$\frac{\partial L}{\partial \theta_i} = \sum_{j=1}^{\infty} \frac{R_j}{P_j} \frac{\partial P_j}{\partial \theta_i} + \frac{S}{(1 - P_0)} \frac{\partial P_0}{\partial \theta_i}, \quad i = 1, 2, \ldots.$$

It is fairly straightforward to show, integrating by parts in the integral for P_j, that

$$\frac{\partial P_j}{\partial \theta_1} = jP_j - (j+1)P_{j+1}$$

$$\frac{\partial P_j}{\partial \theta_2} = \frac{1}{2}\{j^2 P_j - (j+1)(2j+1)P_{j+1} + (j+1)(j+2)P_{j+2}\}.$$

Hence, to find $\partial L/\partial \theta_i$ and its derivatives, one only needs to evaluate a finite number of terms of the distribution $P_j(\xi, \sigma^2)$. However, these probabilities are themselves rather difficult to evaluate, at least when it comes to small values of j. It is therefore necessary to adopt numerical integration. For that purpose it is convenient to write P_j in the form

$$P_j = \frac{e^{j\xi + (1/2)j\sigma^2}(2\pi\sigma^2)^{-1/2}}{j!} \int_{-\infty}^{\infty} e^{-e^y}e^{(j-\xi-j\sigma^2)^2/2\sigma^2}dy.$$

The first two terms, P_0 and P_1, were tabulated by Grundy (1951); more complete tables were given by Brown and Holgate (1971). For large values of j ($j \geqslant 10$), Bulmer derived the approximation

$$P_j \approx \frac{(2\pi\sigma^2)^{-1/2}}{j} e^{-(\ln j - \xi)^2/2\sigma^2} \left[1 + \frac{1}{2j\sigma^2} \left\{ \frac{(\ln j - \xi)^2}{\sigma^2} + \ln j - \xi - 1 \right\} \right].$$

For evaluation of the information matrix, the reader is referred to Bulmer (1974), who also wrote a computer program to carry out these calculations.

An advantage of the lognormal model is that s is always finite and can be estimated from the sample. Bulmer proposed the estimator

$$\hat{s} = S/[1 - P_0(\hat{\xi}, \hat{\sigma}^2)],$$

where $\hat{\xi}$ and $\hat{\sigma}^2$ are the maximum likelihood estimators of ξ and σ^2.

The population parameter $\xi_0 = \xi - \ln \mu$ is naturally estimated by $\hat{\xi}_0 = \hat{\xi} - \ln N$. To find the variance of $\hat{\xi}_0$, we should have to consider $\text{cov}(\hat{\xi}, N)$, which is not easily derived for this model.†

Bearing in mind that $se^{\xi_0 + (1/2)\sigma^2} = 1$, we obtain an alternative estimator of s as

$$\hat{s} = e^{-\xi_0 - (1/2)\sigma^2} = Ne^{-\xi - (1/2)\sigma^2}.$$

3.7. Zipf's model

Zipf (1932) proposed that R_j should approximately follow the sequence $j^{-\xi}, \xi > 0$, where ξ was often taken as 2. To avoid the possibility that $E(N) = \infty$, one may introduce a convergent factor writing

$$E(R_j) = \frac{\beta x^j}{j^\delta}, j = 1, 2, \ldots, 0 < x < 1 \qquad (3.20)$$

so that the case $\delta = 1$ reduces to Fisher's model. Good (1953)

This is analogous to the problem of estimating α in the negative binomial model. In that case $\hat{\omega}$ and $\hat{\kappa}$ were the outputs of maximum likelihood theory, while $\hat{\alpha} = N\hat{\omega}/(1 - \hat{\omega})$ is naturally chosen as an estimate of α. Hence, in order to find $\text{var}(\hat{\alpha})$, we have to pay attention to the covariance between N and $\hat{\omega}$.

proposed the revised version

$$E(R_j) = \frac{\beta x^j}{j(j+1)}, \quad j = 1, 2, \ldots, \tag{3.21}$$

mainly because, as he pointed out, it fitted well with some of Zipf's linguistic data. The case $x = 1$ now corresponds to the λ_is possessing a limiting form of the density

$$g(\lambda; \kappa) = \int_0^\infty \frac{\alpha^\kappa}{\Gamma(\kappa)} \lambda^{\kappa-1} e^{-\alpha\lambda} \frac{1}{(\alpha+1)^2} d\alpha.$$

Under the usual assumption of Poisson sampling, we get the distribution

$$P_j = \kappa \frac{\Gamma(j+\kappa)}{\Gamma(j+\kappa+2)}, \quad j = 1, 2, \ldots.$$

Considering now, as we also did with respect to Fisher's model,

$$\lim_{\kappa \to 0} \frac{P_j}{1-P_0} = \frac{1}{j(j+1)},$$

we arrive at the desired sequence, though this is not quite satisfactory, since $E(N) = \infty$. However, it is a simple matter to show that both (3.20) and (3.21) correspond to certain mixtures of log-series distributions.
Write

$$Q_j = (-\ln \omega)^{-1} \frac{(1-\omega)^j}{j}, \quad j = 1, 2, \ldots.$$

Now consider the mixture on ω where ω has the density

$$f_\omega(\omega; \varepsilon) = \begin{cases} \dfrac{-\ln \omega}{1 - \varepsilon + \varepsilon \ln \varepsilon} & \text{for } \varepsilon \leqslant \omega \leqslant 1 \\ 0 & \text{otherwise} \end{cases}$$

giving the distribution

$$P_j = \frac{(1-\varepsilon)}{(1 - \varepsilon + \ln \varepsilon)} \frac{(1-\varepsilon)^j}{j(j+1)},$$

in agreement with (3.21). Though formally correct, this operation is not entirely satisfactory, since we have in fact mixed *after* trunctation and hence carried out some kind of 'conditional mixing'.

Consider instead

$$E(R_j|\alpha) = \alpha \frac{\left(\dfrac{\mu}{\alpha + \mu}\right)^j}{j}, j = 1, 2, \ldots$$

and let α have the density

$$f_\alpha(\alpha) = \begin{cases} \dfrac{1}{K}\alpha^{-1}\left(\dfrac{\mu}{\alpha + \mu}\right)^2 & \text{for } \alpha > \alpha_0 \\[2ex] 0 & \text{otherwise} \end{cases}$$

where

$$K = \ln(\alpha_0 + \mu) - \ln \alpha_0 - \left(\frac{\mu}{\alpha_0 + \mu}\right).$$

Then unconditionally

$$E(R_j) = \frac{\mu}{K}\frac{x^j}{j(j+1)}, j = 1, 2, \ldots,$$

where $x = \mu/(\alpha_0 + \mu)$. Some algebra gives

$$E(S) = \Sigma E(R_j) = \frac{\mu}{k}\left[1 + \frac{(1-x)\ln(1-x)}{x}\right],$$

and for large values of μ we have approximately

$$E(S) = \frac{\mu}{\ln \mu - \ln \alpha_0 - 1}.$$

Note that we made the distribution of α dependent on μ, which will in practice be hard to accept.

Similarily we may arrive at (3.20) writing

$$E(R_j|\alpha) = \frac{\mu\omega}{1-\omega}\frac{(1-\omega)^j}{j}, j = 1, 2, \ldots$$

where $\omega = \alpha/(\alpha + \mu)$, and using

$$f_\omega(\omega) = \begin{cases} \dfrac{1}{-\ln \varepsilon}\dfrac{1}{\omega} & \varepsilon \leqslant \omega \leqslant 1 \\[2ex] 0 & 0 \leqslant \omega \leqslant \varepsilon, \end{cases}$$

giving

$$E(R_j) = \frac{\mu(1 - \varepsilon)^j}{j^2(-\ln \varepsilon)},$$

in agreement with (3.20). Note again that the corresponding density of α depends on μ. Since ε in practice will be quite small, we have approximately

$$E(S) \approx \sum_{j=1}^{\infty} \frac{\mu}{j^2(-\ln \varepsilon)} = \frac{\pi^2 \mu}{6(-\ln \varepsilon)}.$$

As a consequence, an estimator of ε based on N and S is

$$\hat{\hat{\varepsilon}} = e^{-\pi^2 N/(6S)}.$$

3.8. Some other models

We shall consider three models that are all related to the negative binomial. MacArthur's broken stick model (MacArthur, 1957) is formally the special case $\kappa = 1$ of the negative binomial. It can be generated in the following way: Suppose that $(s - 1)$ points are thrown at random on a line segment of unit length. Let $p_1, p_2, \ldots, p_s$ be the lengths of the segments into which the line is partitioned. Then, any partition is equally likely, and $p_1, p_2, \ldots, p_s$ must be uniformly distributed over the space given by $\sum_{i=1}^{s} p_i = 1, p_i \geq 0$, $i = 2, \ldots, s$. But this is equivalent to the Dirichlet distribution with parameter $\kappa = 1$. In this case, the number of classes equals the parameter α. For further discussion of this model, the reader is referred to Pielou (1969) and Webb (1974).

The negative binomial series can be generalized as indicated in Section 3.4, assuming that the λ_i possess the distribution

$$\frac{\alpha^\kappa}{\Gamma(\kappa)} \frac{1}{(1 - \varepsilon^\kappa)} (1 - e^{-\beta x}) x^{\kappa - 1} e^{-\alpha x},$$

where $\varepsilon = \alpha/(\alpha + \beta)$, $\kappa > -1$ and $\beta > 0$. In Section 3.4 we derived the negative binomial with $\kappa > -1$ by considering the limit $\beta \to \infty$. Further, compounding with the Poisson distribution we get

$$E(R_j) = \frac{\alpha \omega^{\kappa - 1} \Gamma(\kappa + j)(1 - \omega)^j(1 - \eta^{\kappa + j})}{(1 - \varepsilon^{\kappa + 1})\Gamma(\kappa + j)j!}, j = 0, 1, \ldots,$$

where $\eta = (\alpha + \mu)/(\alpha + \mu + \beta)$. It is mainly the first few terms, say $E(R_0)$ and $E(R_1)$, that will differ from a negative binomial series, since $\eta^{\kappa + j}$ soon approaches zero as j gets large. Various estimation methods for this model were considered by Engen (1975b). The number of classes is always finite for this model and can be estimated from the sample. In fact, it is fairly straight forward to show that

$$s = \begin{cases} \dfrac{\alpha(1 - \varepsilon^{\kappa})}{\kappa(1 - \varepsilon^{\kappa + 1})} & \text{for} \quad \kappa > -1, \quad \kappa \neq 0 \\[3mm] \dfrac{-\alpha \ln \varepsilon}{1 - \varepsilon} & \text{for} \quad \kappa = 0. \end{cases}$$

Kempton (1975) proposed that the λ_i possessed the beta distribution of the second kind. This may be viewed as yet another generalization of the negative binomial, since the beta distribution of the second kind is, in fact, a mixture of gamma distributions:

$$g(\lambda) = \int_0^{\infty} \frac{x^{\kappa} \lambda^{\kappa - 1}}{\Gamma(\kappa)} e^{-\lambda x} \frac{\beta^{-q} x^{q - 1}}{\Gamma(q)} e^{-\alpha/\beta} dx$$

$$= \frac{1}{B(\kappa, q)} \frac{\beta^{\kappa} \lambda^{\kappa - 1}}{(1 + \beta\lambda)^{\kappa + q}}.$$

This distribution combined with the Poisson distribution Kempton called the full beta model, but he also proposed considering the limit $\kappa \to 0, s \to \infty$, obtaining what he called the generalized log-series model, since it is a two-parameter model with Fisher's model as a special case. It was shown that this model fitted well with data indicating an exceptionally long tail (to the right). In order to fit the model to the data, a general optimizing computer routine (Ross, 1970) was applied to the grouped observations with class intervals approximately equal on a logarithmic scale.

3.9. Some concluding remarks

We have seen that a number of competing abundance models have been proposed in the literature. In Part II of this monograph an attempt will be made to demonstrate that quite contradictory theoretical models can receive support from the same observational data. We are therefore left with a difficult choice. The following three guidelines may be useful when it comes to choosing amongst a number of competing models.

(1) In the first place we should make sure that the theory applied is logically consistent. In particular we should examine the model with a view to finding out how the population parameters are in theory expected to vary with the sample size.

(2) The theoretical assumptions made should be realistic with respect to the real situation under consideration.

(3) The acceptability of a model is reinforced if there is some realistic process that generates it. By 'realistic' we mean that the process must be in agreement with established scientific facts relating to the matter in hand.

The models dealt with in this chapter are all logically consistent. This does not apply to all diversity models proposed in the literature, however. It has been argued quite often that the sequence $\{R_j\}_{j=1}^{\infty}$ should have some specified smooth mathematical form; often no attempt has been made to show that there exists a population with a structure supporting this assumption. The second guideline above is an important one and will be discussed in detail at the beginning of Part II. At this stage we shall only return for a moment to Zipf's model, in order to demonstrate the link between (1) and (2). Consider the sequence $E(R_j) = \beta x^j/j^2, j = 1, 2, \ldots$. We have seen that there is a population (a mixture of log-series models) which is in agreement with the above series. Hence, we have not departed from (1). But in order to follow (1), we made the distribution the population parameter α dependent on μ. A corollary to this is that the distribution of $\mathbf{p} = (p_1, p_2, \ldots)$ depends on μ. This is surely not realistic in the linguistic application, and (2) is therefore broken. Note that there may still be other theoretical explanations, which are also in agreement with (2). To the author's knowledge, however, such explanations still remain to be found.

Guideline (3) definitely touches a field of great scientific interest, which, however, is extremely resistant to systematic examination. The reason why it is so difficult for some to grip the problems mentioned in (3) is that the abundances appearing in the process can hardly be regarded as independent random variables, except perhaps in a few cases. It is, for example, hard to imagine an animal population evolving without some sort of competition between species. Processes that give a mathematical (or statistical) explanation of the observed structure without taking into account interaction between abundances will usually be unrealistic and therefore of academic interest only.

Some scientists will perhaps go so far as to say that the models considered in this chapter are quite useless in the absence of some generating processes supporting them. In the author's view, this objection is not quite valid. Having said this, however, I would like to add that I nevertheless consider research in the field of stochastic-process and time-series analysis as a task of paramount importance in this context. As to the applicability of already existing class frequency models, however, I prefer to approach the problem from the opposite direction: because we have little quantitative knowledge as to how the population structure has evolved, the present models are applicable as a first aid in our endeavour to understand the nature of many populations; our knowledge of populations with a large number of classes and interacting abundances is likely to remain incomplete for some foreseeable future. Further, it seems advisable to establish what the population structure is before putting too much effort into the much more difficult task of explaining its evolution.

References to some processes generating the log-series distribution were given in Section 3.3. Some of these processes are formulated as birth and death processes. Karlin and McGregor (1967) discussed a model where new species (classes) were supposed to arise as a Poisson process in time, and the descendants to evolve in time according to linear birth and death processes. Kendall (1948) and Wette (1959) considered birth and death processes leading to the negative binomial series. Interaction between species was not taken into account in any of these papers.

Processes supporting the lognormal model are essentially based on the central limit theorem (MacArthur, 1960; Pielou, 1969, 1975; Bulmer, 1974), either (a) by explaining why the λ_i are expected to be log-normally distributed, or (b) by carrying out a sequential breakage process on a given line segment. In the case of (b), the p_i should approximately possess the lognormal distribution. This last point of view is a generalization of MacArthur's broken stick model (see Section 3.8.—and Pielou, 1975). Other generalizations of this breakage process leading approximately to the negative binomial (with $\kappa > -1$) have been proposed by the author (Engen, 1975).

CHAPTER FOUR

Sample coverage

4.1. Introduction

One of the main results of Good's Bayesian approach dealt with in section 2.7 stated that $E(R_1)/N$ was approximately the same as the expected total abundance of those classes that were not represented by any element in the sample. Take as an example Eldrige's statistics of fully inflected words in American newspaper English (Eldridge, 1911; Zipf, 1932; Good, 1953). A sample of $N = 43\,989$ (elements) words and $S = 6001$ different words (observed numbers of classes) was taken. The first terms of the sequence $\{R_j\}_{j=1}^{\infty}$ were, 2976, 1079, 516, 294, 212, ..., giving $R_1/N = 0.068$. Consequently, a foreigner who learned all 6001 words which occurred in the sample would afterwards meet new words in about 6.8% of the words he read. This is a very high percentage, taking into account the large number of words sampled. The initial (prior) probability that none of the words that did not appear in the sample should be observed is actually as small as $(1 - 0.068)^{43\,989} \approx 10^{-914}$. This number may at first glance look suspiciously small, but it is well known in probability theory that quite unlikely events do occur.

Let us suppose that all words in English were listed in advance. Clearly, the above sample falls far short of covering the whole list of words. Some theory that not only defines the 'coverage' of a sample but also provides methods of estimation is called for. Though linguistic data are extreme in the sense that the number of classes is very large, linguistics is by no means the only field of application for such a theory. We shall see that animal populations often fit reasonably well with the negative binomial series with κ as small as, say -0.4. Then $R_1 \approx \alpha(\alpha/(\alpha + N))^{\kappa}$ is actually expected to increase with the sample size. For Fisher's model R_1 is expected to be about constant. Engen (1974) examined some insect data of Mechninick (1964) with $N = 2220$ and $R_1 = 50$, giving $R_1/N = 0.023$.

Consequently, by doubling the sample size, one would expect that about 50 individuals are of previously undiscovered species.

Let $f(p)$ be some positive function of p. Engen (1975c) defined the coverage of the sample with respect to f, $C(f)$, as follows. Write

$$\sum_{i=1}^{s} f(p_i) = \Sigma_I f(p_i) + \Sigma_{II} f(p_i),$$

or equivalently

$$\Phi(f) = \Phi_1(f) + \Phi_2(f),$$

where Σ_I is taken over all classes represented in the sample and Σ_{II} over all undiscovered classes. Then

$$C(f) = \Phi_1(f)/\Phi(f). \tag{4.1}$$

Note that $C(f)$ is a random variable, though it is not observable. We shall here mainly be concerned with its expected value, but the problem of drawing inference about $C(f)$ itself will also be dealt with.

4.2. Finite populations

In this section we adopt the notation used in Section 2.3. There are x_j elements of the class C_j in the population; r_i is the number of x_j's that is exactly equal to i. Capital letters X_j and R_i denote the analogous variables for the sample. As to the number of elements belonging to classes not revealed by the sample, we may adopt the terminology of the previous section, writing

$$U = [1 - C(p)]n.$$

Now

$$E(U) = \sum_{j=1}^{s} \frac{x_j \binom{n - x_j}{N}}{\binom{n}{N}} \tag{4.2}$$

and

$$E(R_j) = \sum_{i=1}^{s} Pr(j \mid x_i, n, N) = \sum_{i=1}^{n} r_i Pr(j \mid i, n, N), \tag{4.3}$$

where

$$Pr(j \mid i, n, N) = \frac{\binom{i}{j}\binom{n-i}{N-j}}{\binom{n}{N}}.$$

In order to express $E(U)$ by the $E(R_j)$ we need the following lemma.
Lemma. Write

$$Q_i(x, n, N) = \frac{Pr(i \mid x, n, N)}{\binom{N}{i}}.$$

Then, if $N \geq i + 1$,

$$Q_i(x, n, N)(x - i) = Q_{i+1}(x, n, N)(n - N) - Q_{i+1}(x, n, N)(x - i - 1).$$

We omit the proof which is just a matter of inserting the definition.
Now, the terms of the sum (4.2) are of the form

$$\frac{x\binom{n-x}{N}}{\binom{n}{N}} = \frac{\binom{x}{1}\binom{n-x}{N-1}(n - x - N + 1)}{\binom{n}{N}\binom{N}{1}},$$

which equals

$$Q_1(x, n, N)(n - N) - Q_1(x, n, N)(x - 1).$$

Using the lemma inductively on the last term we obtain

$$E(U) = (n - N) \sum_{j=1}^{s} \sum_{i=1}^{N} Q_i(x_j, n, N)(- 1)^{i+1}$$
$$+ \sum_{j=1}^{s} Q_N(x_j, n, N)(x_j - N)(- 1)^{N}.$$

Alternatively, this may be expressed as

$$(n - N) \sum_{i=1}^{N} \sum_{j=1}^{n} r_j Q_i(j, n, N)(- 1)^{i+1}$$
$$+ \sum_{j=N}^{n} r_j Q_N(j, n, N)(j - N)(- 1)^{N}.$$

Comparing this with (4.3) we see at once that

$$E(U) = (n - N) \sum_{i=1}^{N} \frac{E(R_i)}{\binom{N}{i}}(-1)^{i+1} + (-1)^{N} \sum_{i=N+1}^{n} r_i \frac{\binom{i}{N}}{\binom{n}{N}}(i - N).$$

$$(4.4)$$

Now if $r_i = 0$ for $i > N$, the last sum vanishes, and

$$E[C(p)] = 1 - \frac{n - N}{n} \sum_{i=1}^{N} \frac{E(R_i)}{\binom{N}{i}}(-1)^{i+1}. \qquad (4.5)$$

It follows from the lemma in Theorem 3.1 that

$$\hat{C}(p) = 1 - \frac{n - N}{n} \sum_{i=1}^{N} \frac{R_i}{\binom{N}{i}}(-1)^{i+1}, \qquad (4.6)$$

under the above assumption is the only existing unbiased estimator of $E[C(p)]$. Note that if N is not too small, the first term in the sum of (4.6) will be large compared with the others, giving the simple estimator

$$\hat{\hat{C}}(p) = 1 - \frac{n - N}{n} \frac{R_1}{N}. \qquad (4.7)$$

We see that $\hat{\hat{C}}(p)$ is effectively the same as Good's result with a factor $(n - N)/n$ correcting for the fact that the population is finite.

Note that there is a strong restriction that must be satisfied in order that $\hat{C}(p)$ should be unbiased. In fact, the sample size must not be smaller than the maximum number of elements contained in any class of the population. Otherwise there will be a bias, which is exactly the last sum of (4.4) divided by n. However, the bias is likely to be small compared with $C(p)$ and $1 - C(p)$ in many practical situations.

4.3. Infinite populations

If we let $n, x_i \to \infty$ so that $x_i/n \to p_i$, we arrive at the multinomial model. It is a simple matter to check that the bias term of $\hat{C}(p)$ now tends to $(-1)^N \sum_{i=1}^{s} p_i^{N+1}$, which can often be neglected compared with $C(p)$ and $1 - C(p)$. Some care must, however, be shown if one of the p_i is close to 1. The simplified estimator now takes the even simpler form

$$\hat{\hat{C}}(p) = 1 - \frac{R_1}{N} \tag{4.8}$$

which is actually the same as Good's estimator.

In order to draw any inference concerning $C(p)$, which interests us a good deal more than $E[C(p)]$, it seems appropriate to consider the distribution of $C(p) - \hat{C}(p)$, taking into account that $C(p)$ and $\hat{\hat{C}}(p)$ are both random variables, possibly correlated. By writing $C(p) - \hat{\hat{C}}(p)$ as a sum of contributions from each class, it is a simple matter to show that the following approximations are valid for the first three cumulants of $C(p) - \hat{\hat{C}}(p)$:

$$K_1 \approx 0, \quad K_2 \approx [E(R_1) + 2E(R_2)]/N^2, \quad K_3 \approx [E(R_1) - 6E(R_3)]/N^3.$$

This enables us relatively simply to find some approximate confidence intervals for $C(p)$, such as the 95% interval

$$1 - \frac{R_1 - 2\sqrt{(R_1' + 2R_2')}}{N} < C(p) < 1 - \frac{R_1 + 2\sqrt{(R_1' + 2R_2')}}{N}. \tag{4.9}$$

Here $R_1', R_2', \ldots$ denotes a smoothed version of the sequence $\{R_i\}_{i=1}^{\infty}$ (see Good, 1953).

Applied to the Eldridge statistics considered in Section 4.1, we get the 95% interval $[0.9291, 0.9355]$. The skewness of $C(p) - \hat{\hat{C}}(p)$, i.e. $K_3/K_2^{3/2}$, is estimated to

$$\frac{R_1 - 6R_3}{(R_1 + 2R_2)^{3/2}} = -0.000\,33,$$

indicating that the normal approximation is likely to work well in this case.

4.4. Some results for the negative binomial model

4.4.1. *Confidence interval for $C(p)$*

For the pseudo-moment method (see Section 3.4) the estimation equations for ω and k were

$$\frac{N}{S} e^{T\hat{\omega} \ln \hat{\omega}/(1-\hat{\omega})} = 1 - \hat{\omega}(T - N/S + 1), \tag{4.10}$$

$$\hat{\kappa} = T\hat{\omega}/(1 - \hat{\omega}) - 1, \tag{4.11}$$

where $T = \Sigma(R_i)i(i - 1)$.

In case the negative binomial fits well, we can derive confidence intervals for $C(p)$ that are expected to be shorter than those given by (4.9). By writing $C(p), S, N$ and T as sums of contributions from the various classes and approximating sums by integrals, one finds

$$E[C(p)] \approx 1 - \omega^{k+1}, \tag{4.12}$$

$$\text{var}[C(p)] \approx \frac{k+1}{\alpha}\left[\omega^{k+1} - \left(\frac{\omega}{2-\omega}\right)^{k+2}\right], \tag{4.13}$$

$$\text{cov}[T, C(p)] \approx \omega^{k+1}(1 - \omega)^2(k + 1)(k + 2), \tag{4.14}$$

$$\text{cov}[S, C(p)] \approx \omega^{k+1} - \left(\frac{\omega}{2-\omega}\right)^{k+1}, \tag{4.15}$$

$$\text{cov}[N, C(p)] \approx \omega^{k+1}(1 - \omega)(k + 1) \tag{4.16}$$

(for the derivation of (4.14) as an example, see Engen, 1975c). The expected coverage can now be estimated by $\hat{C}(p) = 1 - \hat{\omega}^{k+1}$, and we need to evaluate an approximation for

$$\sigma^2(p) = \text{var}[\hat{C}(p) - C(p)|N].$$

A first order approximation may be found by expanding (4.10) and (4.11), using equations (4.13)–(4.16). Fortunately, $N\sigma^2(p)$ turns out to be a function of k and ω only. Tables for $\sqrt{N}\sigma(p)$ were given by Engen (1975c) and are reproduced in Appendix B.

For the data of Mehninick (1964) already mentioned, we have $R_1 = 50$, $R_2 = 20$, $R_3 = 11$, $\hat{\kappa} = -0.366$, $\hat{\alpha} = 5.81$, $N = 2220$. $\hat{C}(p)$ is calculated to 0.9770 and from the tables of Engen (1975c) we find $\sigma(p) \approx 0.0035$. It follows, by using the normal approximation, that the 95% confidence interval is approximately [0.970, 0.984]. If we use (4.9), the interval turns out to be [0.968, 0.986] which is about 30% longer.

4.4.2. Confidence intervals for $C(-p \ln p)$

Writing $f(p) = -p \ln p, \Phi(f)$ is the information index of diversity, and $\Phi_1(f)$ is the contribution to it from the observed classes. As in the previous section we find

$$\Phi \approx \ln \alpha - \psi(k + 1), \tag{4.17}$$

where $\psi(\cdot)$ is the digamma function. Further,

$$E(\Phi - \Phi_1) \approx \omega^{k+1}[\ln(\alpha + v) - \psi(k+1)] \qquad (4.18)$$

and the coverage is estimated by

$$\hat{C}(-p\ln p) \approx 1 - \hat{\omega}^{k+1} \frac{\ln(\hat{\alpha} + N) - \psi(\hat{\kappa} + 1)}{\ln\hat{\alpha} - \psi(\hat{\kappa} + 1)}. \qquad (4.19)$$

First order approximation to

$$\sigma^2(-p\ln p) = \text{var}\left[\hat{C}(-p\ln p) - C(-p\ln p)\big| N\right]$$

as a function of α, k and N can be found as in the previous section. The necessary results analogous to (4.13)–(4.16) were listed by Engen (1975c), who also tabulated $\sigma(-p\ln p)$. The tables are reproduced in Appendix B.

For the data dealt with in Section 4.4.1 we find $\hat{C}(-p\ln p) = 0.9341$ and from table, $\sigma(-p\ln p) \approx 0.0090$. This gives the 95% confidence interval $[0.9265, 0, 9517]$

CHAPTER FIVE

Indices of diversity and equitability

5.1. Introduction

We mentioned in Section 1.1 that the general aim of an abundance model is to make a study of the whole set of relative abundances in cases where the number of classes is large. In particular, some real valued functions of **p** may be of interest, or in the case of a finite population, some function of $\mathbf{r} = (r_1, r_2, \dots, r_n)$. It will be recalled that r_i is the number of classes with i representatives in the population. It was shown in Theorems 2.1 and 2.2 that there are considerable difficulties involved in the estimation of certain functions. This, we argued, was one purely theoretical reason for introducing class frequency models. In this chapter, indices of diversity and equitability will be discussed, and an attempt will be made to examine their relationship to parameters in abundance models. Consequently, the estimation theory adduced in Chapter 3 can be utilized in the estimation of the indices mentioned above.

5.2. Finite populations

The most common measure of diversity for a finite collection of elements is the Brillouin index (Brillouin, 1962)

$$H_b = n^{-1} \ln \left(\frac{n!}{x_1! x_2! \dots x_s!} \right), \tag{5.1}$$

Where the notation is as in Chapter 1. Alternatively, this may be written in the form of equation (1.2). It was shown in Theorem 2.1 that if the sample size N is not smaller than $q = \max_i (x_i)$, then there exists one and only one unbiased estimator of $r_1, r_2, \dots, r_q$, say $\hat{r}_1, \hat{r}_2, \dots, \hat{r}_q$. Consequently, by the lemma of Theorem 2.1,

$$\hat{H}_b = \sum_{i=1}^{n} \left[\frac{\ln i - \hat{r}_i \ln (i!)}{n} \right] \tag{5.2}$$

is the only existing unbiased estimator of H_b when $N \geq q$. The explicit expression for $\hat{r}_i$ was given in Theorem 2.1. Further, if $N < q$, no unbiased estimator exists. However, the Brillouin index is not likely to be used in practice unless the population has been fully censused (Pielou, 1969).

Another measure of diversity used for finite populations is the Simpson index, also defined in Section 1.3.3. This was originally proposed for the multinomial model (Simpson, 1949) as the probability that two randomly chosen elements belong to different classes. For the finite model this is

$$H_s = 1 - \sum_{i=1}^{s} \frac{x_i(x_i - 1)}{n(n - 1)} = 1 - \sum_{i=1}^{n} \frac{i(i - 1)r_i}{n(n - 1)}. \tag{5.4}$$

$$H_s = 1 - \sum_{i=1}^{N} \frac{i(i - 1)R_i}{N(N - 1)} \tag{5.5}$$

is generally an unbiased estimator of H_s. Finally, the number of classes s can be estimated by

$$\hat{s} = \sum_{i=1}^{N} \hat{r}_i \quad \text{(Goodman, 1951)},$$

which is unbiased if $N \geq q$. Otherwise no unbiased estimator exists.

5.3. Infinite populations

5.3.1. *Indices of diversity*

Good (1953) proposed

$$H_{i,j} = \sum_{l=1}^{s} p_l^i(-\ln p_l)^j \tag{5.6}$$

as a class of diversity measures. The most common measures of diversity are actually expressible by the $h_{i,j}$. Simpson's index takes the form

$$H_s = 1 - \Sigma p_l^2 = 1 - H_{2,0}, \tag{5.7}$$

while the information index is

$$H = \sum_l p_l(-\ln p_l) = H_{1,1}. \tag{5.8}$$

The number of classes too, which is sometimes used as a measure of diversity, is within Good's framework expressible as $h_{0,0}$.

Many other measures of diversity have been proposed in the literature, but in the subsequent paragraphs we shall concentrate on the above functions of **p**. Let us just mention another family of measures, say

$$J_t = \frac{\ln(\Sigma p_i^t)}{1-t} = \frac{\ln(H_{t,0})}{1-t}$$

which in information theory is known as the entropy of order t. In particular $J_1 = \lim_{t \to 1} J_t = H_i$, and $J_2 = -\ln(1 - H_s)$ (see Renyi, 1961; Pielou, 1975).

Indices of (ecological) equitability

The indices of diversity presented in the previous section all have the properties (1) & (2) listed in Section 1.3.2. Attempts have been made in ecological literature to define indices separating these two features. Measures of equitability are functions of **p** that are not much affected by the value of s, but increase as the abundances tend to be more equal. The concept of equitability was introduced by Lloyd and Ghelardi (1964), who proposed the following measure: write s' for the number of classes required in the broken stick model to give the same value of H as for the population under consideration. Then the equitability is defined as $\epsilon = s'/s$. The broken stick model is chosen as a 'yardstick' because it seems (approximately at least) to represent the maximum equitability for real biological communities. It follows that $\epsilon = 1$ for a population fitting the broken stick model, and $\epsilon = 0$ for one fitting the logarithmic series; this is because the latter model represents a population with an infinite number of species but a finite value of H (see Section 5.4) and s'. Pielou (1969) used the index $H/\max(H) = H/\ln s$, and e^H/s has also been proposed (Buzas and Gibson, 1969; Sheldon, 1969). The above indices of equitability, however, do not seem to be appropriate in general. Take, for example, Lloyd and Ghelardi's index. For the negative binomial model the value of ϵ coincides with the value of the shape parameter κ for $\kappa = 1$ and $\kappa = 0$. The disadvantage lies in the fact that ϵ, as well as the other measures mentioned above, will be zero for all negative values of κ. Intuitively, equitability should increase with κ also in this region, since, as κ gets smaller, rare classes get more rare and the common ones more abundant. The above measures do not account for this, and since negative values of κ are likely to occur in practice, none of them seem to be appropriate in general. In order to define a measure that answers the purpose, let us consider the random variable Z, with distribution

$$Pr(Z = -\ln p_j) = p_j, \qquad j = 1, 2, \ldots, s,$$

where s may be infinite. Now $E(Z) = -\Sigma p_j \ln p_j = H$. A natural choice of an *inverse* measure of equitability now seems to be

$$H_v = \mathrm{var}\, Z = \Sigma p_j (\ln p_j)^2 - (\Sigma p_j \ln p_j)^2 \qquad (5.9)$$

We shall call H_v the index of variability (Engen, 1977a). H_v is expressible by Good's indices as

$$H_v = H_{1,2} - H_{1,1}^2 \qquad (5.10)$$

The variability can only be zero if all the abundances are equal; otherwise $H_v > 0$.

Random populations

All indices considered in Sections 5.3.1 and 5.3.2 are random variables in case **p** is not considered as fixed. However, it is convenient to attach to any population some fixed number (population parameter) describing some of its features independently of the actual realization of **p**. For the indices dealt with in Section 5.3 we shall replace $H_{i,j}$ by its expected value $E(H_{i,j})$. Notice that H_v is not replaced by $E(H_v)$. This would be an inconvenient definition in so far as the bivariate distribution of (p_i, p_j), $i \neq j$, would then have to be examined. Such a complication is quite unnecessary for our purposes. We shall write

$$E'(H_v) = E(H_{1,2}) - [E(H_{1,1})]^2.$$

5.4. Parameter transformation

Introduction

Let t be a real positive number. Write $H_{t,0} = \sum_{l=1}^{s} p_l^t$. Assuming that the jth derivative of $H_{t,0}$ exists, we have

$$\frac{\partial^j H_{t,0}}{\partial t^j} = \sum_{l=1}^{s} p_l^t (\ln p_l)^j. \qquad (5.11)$$

Hence

$$E(H_{i,j}) = (-1)^j \frac{\partial^j E(H_{t,j})}{\partial t^j}\bigg|_{t=i} \tag{5.12}$$

But $E(H_{t,0})$ can be expressed in terms of the structural distribution; in fact

$$H_{t,0} = \sum_{i=1}^{s} p_i^t = \int p^{t-1} h^*(p|\mathbf{p})dp$$

and

$$E(H_{t,0}) = \int p^{t-1} h^*(p)dp \tag{5.13}$$

As a consequence, all the above mentioned indices may be expressed by the structural distribution $h^*(p)$.

The negative binomial model

It was shown in (3.4) that for any $\kappa > -1$ the structural distribution is

$$h^*(p) = B(\kappa + 1, \alpha - \kappa)^{-1} p^\kappa (1-p)^{\alpha-\kappa-1}.$$

This is exact for the random model and an approximation when the abundances are considered fixed. By simple integration

$$E(H_{t,0}) = \frac{\Gamma(\alpha + 1)\Gamma(\kappa + t)}{\Gamma(\kappa + 1)\Gamma(\alpha + t)}, \tag{5.14}$$

giving the following relations:

$$E(H_s) = 1 - \frac{\kappa + 1}{\alpha + 1} \tag{5.15}$$

$$E(H) = \psi(\alpha + 1) - \psi(\kappa + 1) \tag{5.16}$$

$$E'(H_v) = \psi'(\kappa + 1) - \psi'(\alpha + 1) \tag{5.17}$$

where $\psi'(\cdot)$ denotes the trigamma function. The following relations (Abramowitz and Stegun, 1964) are appropriate for the evaluation of $E(H)$ and $E(H_v)$

$$\psi(x) = \ln x - \tfrac{1}{2}x^{-1} - \tfrac{1}{12}x^{-2} + \tfrac{1}{120}x^{-4} - \tfrac{1}{252}x^{-6} + \ldots, x > 1.$$

$$\psi(x + 1) = \psi(x) + \frac{1}{x}.$$

$$\psi'(x) = x^{-1} + \tfrac{1}{2}x^{-2} + \tfrac{1}{6}x^{-3} - \tfrac{1}{30}x^{-5} + \tfrac{1}{42}x^{-7} - \ldots, x > 1.$$

$$\psi'(x+1) = \psi'(x) - \frac{1}{x^2}.$$

The corresponding results for Fisher's logseries model and MacArthur's broken stick model come out as the special cases $\kappa = 0$ and $\kappa = 1$ in relations (5.15), (5.16) and (5.17). Inserting $\psi(1) = -\gamma$, where $\gamma = 0.57772\ldots$ is Euler's constant, $\psi(2) = 1 - \gamma, \psi'(1) = \frac{\pi^2}{6}, \psi'(2) = \frac{\pi^2}{6} - 1$, we get for the logseries model

$$E(H_s) = \frac{\alpha}{\alpha + 1}, \tag{5.18}$$

$$E(H) = \psi(\alpha + 1) + \gamma, \tag{5.19}$$

$$E(H_v) = \frac{\pi^2}{6} - \psi'(\alpha + 1), \tag{5.20}$$

and for MacArthur's model

$$E(H_s) = \frac{s - 1}{s + 1}, \tag{5.21}$$

$$E(H) = \psi(s + 1) + \gamma - 1, \tag{5.22}$$

$$E'(H_v) = \frac{\pi^2}{6} - 1 - \psi'(s + 1), \tag{5.23}$$

recalling that $s = \alpha$ when $\kappa = 1$. Equation (5.15) was derived by Eberhardt (1969); (5.18) by Simpson (1949); (5.16) and (5.19) by Bulmer (1974); and (5.21) and (5.22) by Webb (1974).

Some of the above relations are shown in Fig. 2 graphically.

The geometric series model

The geometric series model dealt with in (3.5) is defined by

$$p_i = \theta(1 - \theta)^{i-1}, i = 1, 2, \ldots, 0 < \theta < 1.$$

Consequently

$$H_{t,0} = \sum_{i=1}^{\infty} \theta^t [(1 - \theta)^t]^{i-1} = \frac{\theta^t}{1 - (1 - \theta)^t}. \tag{5.24}$$

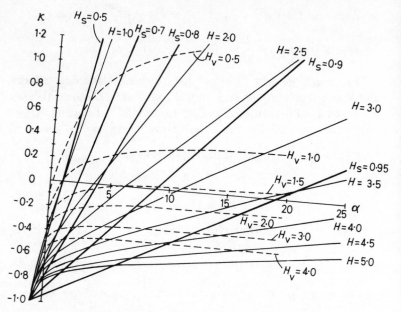

Fig. 2 *Some relations referring to the negative binomial model*

For the Simpson index we find

$$H_s = 1 - H_{2,0} = \frac{2(1-\theta)}{2-\theta} = \frac{\alpha}{\alpha+1/2}, \tag{5.25}$$

where $\alpha = (1-\theta)/\theta$. Further, after some algebra

$$H = -\ln\theta - \frac{(1-\theta)\ln(1-\theta)}{\theta}, \tag{5.26}$$

$$H_v = \frac{(1-\theta)[\ln(1-\theta)]^2}{\theta}. \tag{5.27}$$

The lognormal model

Write $K(t)$ for the cumulant generating function of $\ln p_i$. Then

$$E(H_{t,0}) = se^{K(t)}. \tag{5.28}$$

In Section 3.6 we derived an approximation for $K(t)$ which is asymptotically correct as either $v = \sigma^2 \to 0$ or $s \to \infty$. By inserting this approximation into (5.28) we obtain

$$E(H_s) \approx \frac{1 - s^{-1}}{1 + (e^v - 1)/s},$$ (5.29)

$$E(H) \approx \ln s - \tfrac{1}{2}v + \tfrac{1}{3}\ln\left[1 + (e^v - 1)/s\right],$$ (5.30)

$$E'(H_v) \approx v - \ln\left[1 + (e^v - 1)/s\right], \qquad (v \to 0 \text{ or } s \to \infty).$$ (5.31)

Longuet-Higgins (1972) and Bulmer (1974) used the normal approximation for $\ln p_i$, obtaining

$$E(H) \approx \ln s - \tfrac{1}{2}v$$ (5.32)

I have compared this formula with simulation results, and it seems to break down even in the case of fairly small values of v, though it is asymptotically correct as $v \to 0$. Even the improved formulae (5.29), (5.30) and (5.31) cannot be recommended for $v > 5$, unless s is extremely large.

PART II

Ecological applications

CHAPTER SIX

Abundance models in ecology

6.1. Introduction

As we mentioned in Section 1.1.' there can be little doubt that ecology was responsible for the development of the theory of diversity and that it is still the main field of application. Throughout this chapter, the classes $C_1, C_2, \ldots, C_s$ refer to the various species of some taxonomic group of animals or plants. The possibility of applying the theory to higher order classification, for example by viewing the elements as species and the classes as genera, will be discussed in Chapter 7.

The present monograph is not meant to be a textbook in 'ecological diversity': it is primarily concerned with stochastic abundance models, including some judgement of the realism of the assumptions on which the various models are based. Readers interested in further discussion of diversity models and their applications in the science of ecology are referred to Pielou's book on ecological diversity (Pielou, 1975).

6.2. Some historical remarks

In Part I we chose the vectors of abundances, λ or $\mathbf{p}$, as the starting point for the analysis of diversity. Perhaps this is not the natural thing to do when your background in statistics and model building is poor. To liberate oneself from one's own observations and think in terms of a model still seems to be difficult for many scientists. In view of this it is not surprising that the first 'models' in ecology that bear some relation to the concept of diversity were mathematically describable *properties of samples* from populations with a large number of species. In our notation, properties of $\mathbf{R}$ were examined. It is a striking feature of samples from animal and plant populations that the number of species (S) increases with the number (N) of individuals sampled, even if the samples are quite large. With no statistical tool available for analysing these data, ecologists started by plotting S against N (or the sample area), producing what is known as species–area curves. Arrhenius (1921) found that ln S was approximately proportional to ln N for botanical data. However, Gleason (1922) soon criticised his work, arguing that S was proportional to ln N. As we saw in Part I, these curves definitely reveal properties relating to the concept of diversity. Arrhenius' findings correspond to the negative binomial model with $-1 < \kappa < 0$, while Gleason's formula is equivalent to Fisher's model. Mehninick (1964) proposed that S was proportional to $\sqrt{N}$ for an insect population, which again corresponds to the negative binomial model with $\kappa = -\frac{1}{2}$. With the statistical methods now available in this field, the species–area curves seem to be of minor importance.

The concept of 'sample coverage' dealt with in Chapter 4 is also in a way related to the species–area curve. By plotting the curve, some ecologists held that it was possible to arrive at what they called the 'minimum area'; that is, roughly speaking, they thought they would be able to estimate the minimum sample size required to find 'most of the species' in the population. This problem is equivalent to that of estimating the number of species in the community. However, as was demonstrated by Theorem 2, this task is exceedingly difficult (and sometimes quite impossible) in the absence of a species frequency model. To estimate the number of species by examining the curve by eye is not to be recommended, except perhaps in the extreme case when the slope is zero for quite a long interval, indicating that all the species have been found.

A second ecological approach which is essentially different from ours, is that of plotting the order statistics of the X_i, say $X^{(1)} \geqslant X^{(2)} \geqslant \ldots \geqslant X^{(S)}$ against the order. This has mainly been done in connection with MacArthur's broken stick model (MacArthur, 1957), due to the fact that MacArthur derived the expectations of the order statistics of the p_is for his model, say

$$E[p^{(r)}] = \frac{1}{s} \sum_{i=1}^{r} (s - i + 1)^{-1}.$$

It is clear, on reflection, that this approach is not entirely satisfactory either, since order statistics are often somewhat intractable from a theoretical point of view.

A third way of presenting the data graphically is to plot R_i against i. In an ecological context, this method was first adopted by Willis (1922). Later Willis' work was followed up by Fisher, Corbet and

Williams (1943). Hence, historically, there are three different ways of plotting the data: (a) to plot S against N, (b) to plot $X^{(r)}$ against r, and (c) to plot R_j against j. The three plots have, broadly speaking, the same purpose, viz. to reveal properties of the pattern of abundances. Plot (b) exhibits most clearly the properties relating to the most abundant species, while plot (c) is preferable when properties of the less abundant species are to be visualized. In any case, statistical analysis along the lines suggested in Part I ought to be carried out next.

6.3. Interpretations of fixed and random models in ecology

It is beyond the scope of this monograph to go into the respective merits of the various approaches in ecological research. Here we shall merely concentrate on practical interpretations of terms like 'probability', 'expectation', 'variance' and 'dependence' as used in the various models. These statistical terms will turn out to convey quite different meanings depending on whether for example a fixed or a random symmetric model is applied. Clearly, one cannot hope to draw valid conclusions based on the statistical model until the statistics are correctly interpreted. There are several pitfalls one must try to avoid.

Introductory textbooks in statistics often start off by defining the concept of a random experiment and the space of possible outcomes. To any event that may be the result of the experiment we attach a probability. If the same experiment is repeated over and over again, the relative frequency of an event 'converges' to the probability of that event. It seems necessary to return to these basic terms if we are to see clearly the practical consequences of adopting a particular approach in ecology (Engen, 1977b).

What, then, is our random experiment? It may be tempting to answer that this must be the actual sampling of individuals. Such an answer is not, however, necessarily correct.

Let us first consider the vector of abundances, $\lambda = (\lambda_1, \lambda_2, \dots, \lambda_s)$. What we know for certain is that λ is a function of time, say $\lambda = \lambda(t)$, and that there are stochastic elements determining $\lambda(t)$. In fact, $\lambda(t)$ is a multivariate stochastic process. From this point of view the interpretation of a fixed model is straightforward: in a model with fixed abundances, t is fixed and the aim is to analyse the population structure *at that time*. The fact that $\lambda(t)$ is a result of randomness is kept out of the analysis. It is meaningless to say that this *approach* is wrong, though it may not always be preferable. If we want to investigate the process $\lambda(t)$ from the point of view of a fixed model, we need to take samples at different points of time, say at $t_1, t_2, \ldots, t_n$. These samples and the corresponding estimators constructed are, in terms of a fixed model, independent. However, the main advantage of this approach is that we do not need to make any prior assumptions about the process $\lambda(t)$. Note that if $\hat{\theta}(t)$ is the estimate of some population parameter $\theta(t)$, then, for example, the variance of $\hat{\theta}(t)$ is due to sampling only. Saying that $\hat{\theta}(t)$ is unbiased would mean that $E[\hat{\theta}(t)]$ is equivalent to $\theta(t)$, which from another point of view may be a random variable.

Now let us turn to the interpretation of models where $\lambda(t)$ is a random variable. One possibility is to consider the whole process $\lambda(t)$ (the evolution) as the random experiment. In theory, it is not necessary to take account of more than one realization of the process in order to anlyse its properties. If we observe $\lambda(t_1), \lambda(t_2), \ldots, \lambda(t_n)$, we will, under rather general assumptions, theoretically get the required information about the process as $n \to \infty$. However, $\lambda(t)$ evolves through geological time, while we can make our observations only through a very short time interval. It is therefore unlikely that we can in practice obtain any essential information about this process at all, unless we include pollen analysis and fossil studies or other methods that make possible a study through geological time. It would thus appear that we are forced to rely on rather wild guessing with respect to the distribution of $\lambda(t)$ if this interpretation is adopted.

Alternatively, we might assume that $\lambda(t)$ has reached an equilibrium distribution, and that the random experiment consists in choosing a point of time more or less at random. For a certain sequence $t_1, t_2, \ldots, t_n$, the variables $\lambda(t_1), \lambda(t_2), \ldots, \lambda(t_n)$ may be independently distributed (with the equilibrium distribution) if $t_i - t_{i-1}, i = 2, 3, \ldots, n$ are sufficiently large. This interpretation of random models seems plausible, but there are still pitfalls. It is for example a simple matter to demonstrate that the components of λ cannot be independently and identically distributed. At least this is so when s is large, and in particular when the logarithmic series fits the data. Let us consider two samples, say at t_1 and t_2. Suppose further that the components of λ were independently and identically distributed. Let S_1 and S_2 be the number of species represented in these samples and write $A_i, i = 0, 1$, for the event that one particular species is represented in the sample taken at t_i.

Then, conditionally upon S_1 and S_2, we have, assuming independence

and
$$Pr(A_i) = S_i/s, i = 0,1$$
$$Pr(A_1 \cap A_2) = S_1 S_2/s^2.$$

The expected number of species represented by at least one individual in both samples is therefore $S_1 S_2/s$. For models with an infinite or a very large number of species this is zero or approximately zero. The present author (Engen, 1977b) has given examples from the ecological literature showing that the above statement is contradicted by reality. Further, for Fisher's model $X_i(t_1)$, $X_i(t_2), \ldots, X_i(t_n)$, should be independent observations from the negative binomial distribution with κ slightly larger than zero. In fact

$$\lim_{\kappa \to 0} Pr(X_i(t_j) = 0) = 1$$

under these assumptions, and samples taken at two different points of time would have no species in common with probability 1. Again, this is in disagreement with this type of ecological data. Consider, for example, the data of Dirk (1938), which involves catching moths in light traps at Orono, Maine, 1931–1934. During this period, representatives of 240 species were sampled. Out of these species, 126 were found in all four years and 56 were found in three out of four years. If we choose the symmetric model, our random experiment can be interpreted only as evolution; samples taken at different points of time (in this case each year) or at different locations are definitely statistically dependent since they result from the same evolution. This dependence has often been ignored by authors applying symmetric models in the analysis of species abundance data.

This argument seems to be valid also if the number of species in the population is relatively small. In bird data from Quaker Run Valley (Saunders, 1936), Saunders considers the habitats 'Oak-Hickory: mature' and 'Maple-Beech Hemlock: pastured.' There are 25 species in each sample with 17 species represented in both. The total population in the valley is about 85 species, giving an estimate of 7,4 for the expected number of species represented in both samples under the hypothesis of statistically equivalent species and 'independent habitats'. Therefore, the populations in these two habitats are dependent with respect to the symmetric model with independent components of λ.

The above conclusions are in disagreement with that of Anscombe (1950) who notes:

For example, Williams gives figures for capture of Noctuidae during a period of three months in 1933 at two traps (Fisher et al. (1943)). One trap on a roof-top gave $S = 58$, $N = 1856$, $\hat{\alpha} = 11.37$, $\hat{p} = 163$: the other, in field a quarter of a mile away, gave $S = 40$, $N = 929$, $\hat{\alpha} = 8.51$, $\hat{p} = 109$. Fisher's formula (6.15) for the standard error of either estimate of α gives approximately 0.67, indicating a significant difference between them. Whether or not the relative abundances of the species differed at the two traps would be more efficiently tested by comparing counts of individual species. Formula (6.6.) gives for the standard error of either estimate of α approximately 1.59, against which the observed difference is not significant.

Anscombe's formulae (6.15) and (6.6) refer to the fixed and random symmetric model respectively, and the argument seems to be wrong. Formula (6.6), based on independently identically distributed components of λ, could only be used for this kind of test if the samples were independent. If the two estimates are $\hat{\alpha}_1$ and $\hat{\alpha}_2$, the test is based on calculating the variance of $(\hat{\alpha}_1 - \hat{\alpha}_2)$, var $\hat{\alpha}_1$ + var $\hat{\alpha}_2$ − $2\text{cov}(\hat{\alpha}_1, \hat{\alpha}_2)$. The covariance term, which may possibly be relatively large, was not taken into account. Williams (1943) did not give the original data, but since the two samples were taken at a fairly close distance, one must presume that several species are represented in both samples (which always seems to be the case for this kind of data). As demonstrated above, the samples are then dependent and the covariance term can hardly be neglected.

Cohen's derivation of the broken stick model (Cohen, 1968) was also based on a random model with statistically equivalent species. He assumed that the λ_i were independently distributed with the same exponential distribution. However, Cohen expressed a rather suspicious attitude against these assumptions when he wrote:

The strong assumption of constant parameters for the exponential distribution (the same value for all species—this author's insertion), regardless of the composition of the community, does not seem especially compelling to me, even in a comparison of two environments identical except for the species present. The assumption is not especially attractive because the presence or absence of one species alters the situation for the others, even if 'all else' is equal.

If $\lambda(t)$ has reached an equilibrium distribution, multivariate time-series analysis is applicable. However, remembering that the number of components of λ is large and that they are likely to interact in a complicated way, one should not be too optimistic. It seems advisably first to study the time series $\theta(t)$, where θ is a vector of population parameters with a small number of components (say 1 or 2), such as those dealt with in Chapter 3.

Even if we take into account that, as far as the symmetric model is concerned, there is dependence between samples, we are still left with some residual problems, which to the author's knowledge have not received much attention in the literature. Let λ_i be the expected number of individuals of the species C_i within a sampling area of some prescribed size, and assume for example that λ_i is lognormally distributed. Let $\lambda_i^{(1)}$ denote the value of λ_i for some particular randomly chosen area and let $\lambda_i^{(2)}$ be the value of λ_i in an adjacent area of the same size. The two areas together constitute another area of double size and the abundance defined for this area is $\lambda_i^{(1)} + \lambda_i^{(2)}$. It is trivial to show that this sum cannot be lognormally distributed unless the correlation between $\lambda_i^{(1)}$ and $\lambda_i^{(2)}$ is 1, implying $Pr(\lambda_i^{(1)} = \lambda_i^{(2)}) = 1$. Hence, if the lognormal model is to be generally valid for this population, λ ought to be the same over the entire area. Consequently, there can be no clustering of individuals, though we know that clustering usually occurs in practice (Pielou, 1969). The abundance also had to be constant through time, since 'area' can be replaced by the 'time a trap is working' in the above argument. Otherwise the model is valid only for areas of one particular size and shape (or one particular time of sampling when, for example, a light trap is used). We should have to be very lucky indeed if that turned out to be the very sampling area we had chosen.

The same argument can be adduced against MacArthur's broken stick model because a sum of two exponentially distributed random variables cannot be exponentially distributed unless their correlation is 1. As to the negative binomial model, there seems to be a possibility of defining a spatial model generally valid for any area, but systematic research on this problem in relation to species frequency models has not been carried out.

Let, for example, the whole area under investigation be partitioned into quadrats, say of size ΔA, and let the λ_i for the various quadrats be independently gamma distributed with parameters $[\alpha, \kappa(\mathbf{r})\Delta A]$, where $\mathbf{r}$ denotes the centre of the quadrat. Then in the limit as $\Delta A \to 0$, for any area A the corresponding value of λ_i would be a gamma variate with parameters $(\alpha, \int_A \kappa(\mathbf{r})d\mathbf{r})$. It seems worthwhile to carry out research in this field, taking into account that the λ_i

for areas close to each other are likely to be positively correlated, not independent as described above. Further, it seems necessary to investigate the extent to which the other models can be valid as approximations when the above mentioned correlations are not 1.

Note that if a random model is consistent over some area the corresponding fixed model is also logically consistent, since, for any distribution of λ_i we have defined a fixed set of abundances associated with that distribution. This can be done for any area if the random model is well defined.

6.4 The stability of population parameters

As a further demonstration of the meaning of the terms appearing in the various models, we shall see how the stability of population parameters may be examined, following Engen (1977b). Consider as an example Simpson's index of diversity

$$H_s = 1 - \sum_{i=1}^{s} p_i^2.$$

If $\mathbf{X}$ possesses the multinomial distribution, then the minimum variance unbiased estimator of H_s is

$$\hat{H}_s = 1 - \sum_{j=1}^{s} \frac{X_j(X_j - 1)}{N(N-1)} = 1 - \sum_{j=2}^{\infty} \frac{j(j-1)R_j}{N(N-1)},$$

and Simpson derived

$$\text{var}(H_s | \mathbf{p}) \approx \frac{4}{N} [\Sigma p_i^3 - (\Sigma p_i^2)^2], \qquad (6.1)$$

(the exact expression was also derived) which in our terminology, of course, refers to the fixed model. If one uses a symmetric model, it is usually straightforward to evaluate var (H_s). For the negative binomial model we have, for example,

$$\text{var}(H_s) = \frac{(\kappa + 1)}{(\alpha + 1)} \left[\frac{(\kappa + 2)(\kappa + 3) + (\alpha - \kappa)(\kappa + 1)}{(\alpha + 2)(\alpha + 3)} - \frac{(\kappa + 1)}{(\alpha + 1)} \right]. \qquad (6.2)$$

This would be an inverse measure of the stability of H_s, indicating how much H_s would vary in a sequence of 'independent populations'. for example, the populations present at one particular location during subsequent years. However, our conclusion in the previous section implies that (6.2) is inappropriate as such a measure because the underlying model is unrealistic. The true stability of H_s would be more efficiently analysed by considering a sequence of estimates, $\hat{H}_{s_1}, \hat{H}_{s_2}, \ldots, \hat{H}_{s_n}$, with corresponding expectations $H_{s_1}, H_{s_2}, \ldots, H_{s_n}$.

Let us assume that $H_{s_1}, H_{s_2}, \ldots, H_{s_n}$ are independent observations from some distribution. Note that this is not in disagreement with our conclusions in the previous section. Since we are now considering the variations in H_s, we must of course use a random model, but it is sufficient to assume that each component of λ has reached some equilibrium distribution. It is unnecessary to impose the unrealistic assumption that these are independent and identically distributed. Under these assumptions we have

$$\operatorname{var} \hat{H}_s) = E\left[\operatorname{var}(\hat{H}_s|\mathbf{p})\right] + \operatorname{var}\left[E(H_s|\mathbf{p})\right],$$

and since

$$E\hat{H}_s|\mathbf{p}) = H_s,$$

we have

$$\operatorname{var}(H_s) = \operatorname{var}(\hat{H}_s) - E\left[\operatorname{var}(\hat{H}_s|\mathbf{p})\right]. \tag{6.3}$$

Here $\operatorname{var}(\hat{H}_s)$ may be estimated by the sequence $\hat{H}_{s_1}, \hat{H}_{s_2}, \ldots, \hat{H}_{s_n}$, and $E[\operatorname{var}(\hat{H}_s|\mathbf{p})]$ may be estimated by considering (6.1). As the sample size tends to infinity, $E[\operatorname{var}(\hat{H}_s|\mathbf{p})]$ tends to zero, but for relatively small samples it must be taken into account. In fact, it seems most efficient to work with several small samples rather than a few large ones when the purpose is to analyse this sort of stability.

Equation (6.3) may, of course, be generalized by replacing H_s by any population parameter. Hence, our conclusion is that the real stability of a population parameter may be revealed by evaluating the conditional variance of the estimate as a correction for the sampling error as demonstrated by (6.3).

CHAPTER SEVEN

Abundance models in ecology — examples

7.1. Discussion of assumptions

A generating process

In Chapter 2 we made some general assumptions concerning the distribution of $\mathbf{X} = (X_1, X_2, \ldots, X_s)$ stating that the probability generating function of $\mathbf{X}$ was

$$G_{\mathbf{X}}(\mathbf{z}) = M_\nu(\Sigma p_i z_i - 1), \tag{7.1}$$

where $M_\nu(\ )$ denotes some moment-generating function. We now proceed to closer examination of what this assumption means in practice. We shall let the individuals be symbolized by points in a plane; let Ω denote the part of the plane defining the population to be investigated. The sampling of individuals is carried out by placing some area, say a circle or a quadrat, at random in Ω and counting the number of individuals belonging to the various species in this area.

Let $f(\mathbf{u}), \mathbf{u} \in \Omega$, be some integrable positive real-valued function, where $\mathbf{u}$ denotes a point in the plane. Further, let Δx_i be the number of elements belonging to C_i in some small area denoted $\Delta \mathbf{u}$. We shall assume that

$$Pr(\Delta x_i = 1, \Delta x_j = 0 \text{ for } j \neq i) = p_i f(\mathbf{u}) \Delta \mathbf{u} + O(\Delta \mathbf{u}^2)$$

$$Pr(\Delta x_j = 0 \text{ for } j = 1, 2, \ldots, s) = 1 - f(\mathbf{u}) \Delta \mathbf{u} + O(\Delta \mathbf{u}^2).$$

Now consider some area $\omega \in \Omega$. It follows that $X_1, X_2, \ldots, X_s$ for ω are independent Poisson variates with means $\nu(\omega) p_i, i = 1, 2, \ldots, s$, where

$$\nu(\omega) = \int_\omega f(\mathbf{u}) d\mathbf{u}.$$

If the area ω is placed in Ω in some random way, $v(\omega)$ is a random variable, say with moment-generating function $M_v(\cdot)$, which leads us to equation (7.1).

In the above argument the plane may be replaced by the n-dimensional Euclidean space without altering the conclusion. In particular, we may consider the one-dimensional space and let this symbolize time. An interval on the time axis can then be interpreted as the time a trap is working.

Segregated populations

The above model may at first glance look rather general, since there are few restrictions on the function $f(\mathbf{u})$. We have opened up the possibility of any degree of clustering of individuals. However, in the model under discussion all species tend to be clustered round the same locations, which would in practice have to be interpreted as places where conditions are generally good ($f(\mathbf{u})$ takes large values). We know that this is not always the situation in real biological communities. In fact, populations of various kinds tend to be segregated; an individual is likely to be situated closer to individuals of the same species than others (Gleason, 1922; Williams, 1944; Goodall, 1952; Pielou, 1966). Pielou (1966) showed how this pattern may be investigated statistically when the samples are small ($N \leqslant 4$). She applied the technique to forestry data, demonstrating that natural deaths were due to competition between individuals of the same species rather than to competition between species. Segregation is obviously inconsistent with our general assumption given by equation (7.1). On the other hand, the effect is likely to depend considerably on scaling (sample size, quadrat size, time of sampling and so on). If a population shows segregation for small samples, the effect on large samples may still be quite negligible. The opposite tendency to segregation can also be observed in natural populations, but it is very rare (Pielou, 1966).

The efficiency of traps

It is necessary to append a few general remarks on sampling methods. In practice, it is impossible to construct a method so that each individual, regardless of which species it belongs to, has the same chance of being caught. As sample sizes tend to infinity, the species will not tend to be sampled in proportions

$\mathbf{p} = (p_1, p_2, \ldots, p_s)$, but in some other proportions, say $\mathbf{p}^{(m)} = [p_1^{(m)}, p_2^{(m)}, \ldots, p_s^{(m)}]$, where m refers to the method of sampling. Hopefully, $p_i^{(m)}$ should be approximately equal to p_i, but this is not necessarily satisfied. Hence, for a given method of sampling it is the population structure $\mathbf{p}^{(m)}$ and not $\mathbf{p}$ that we are analysing.

The geographic boundaries of the community

Besides restricting the study to a taxonomic group of animals, the population also has to be limited by geographic boundaries. In some situations, limits can be drawn quite naturally, for example when an island (MacArthur, Recher and Cody, 1966; Williams, 1946), a lake, or a fiord is under investigation. In other cases limits will have to be chosen through biological considerations. This leads us to concepts like alpha-, beta-, and gamma-diversity (Whittaker, 1972) or within-habitat and between-habitat diversity (MacArthur, 1957, 1964, 1965). Alpha-diversity, or within habitat diversity, refers to populations limited geographically to one habitat, while gamma-diversity is the total diversity of a population in a region including several habitats. Roughly, beta-diversity or between-habitat diversity has to do with the difference between these two situations. Whittaker (1972) wrote symbolically $\beta = \gamma/\alpha$, and proposed some specific measures of beta-diversity. Further, he says:

The total, or gamma diversity of a landscape, or geographic area, is a product of the alpha diversity of its communities, and the degree of beta-differentiation among them.

And for the mathematical description:

Since gama samples have the same dimensional characteristics (number of species, relative importance values) as alpha samples, the same measurements of diversity and slope (of the species – area curve) are appropriate.

Between-habitat diversity is strongly connected to what have been called measures of similarity between two populations. Example of these are Jaccard's index (Jaccard, 1902), Johnson and Brinkhurst's index (Johnson and Brinkhurst, 1971) and Williams measure of similarity based on the logarithmic series model (Williams, 1944, 1946).

There does not seem to be any obvious answer to the general question whether alpha or gamma samples are in best agreement with our general mathematical assumptions. The most striking difference between alpha- and gamma-populations presumably lies in the relevance of the different factors that control the community structure (MacArthur, 1964, 1965; MacArthur, Recher and Cody, 1966).

Some concluding remarks

Taking into account both the problem of defining the population and the questions concerning our general assumptions discussed in the previous section, it does not appear unreasonable to claim that the models dealt with in Part I are oversimplified. This would obviously be an argument of great weight. On the other hand, it is clear that a reasonable understanding of more complex systems can only be reached through the process of understanding simpler ones. At this point there is no difference between our approach and the way natural science develops in general.

7.2. Judging goodness of fit

We have seen that species frequency models may be defined in such a way that $R_1, R_2, \ldots, R_N$ possesses the multinomial distribution when the number of species observed is considered fixed. In particular, this would be the case if the λ_i are independently and identically distributed. The same would be true if the abundances of the symmetric model are reordered as described in Section 2.2. In that case the classical χ^2-test for goodness of fit would apply. Suppose we group the R_i by

$$G_l = \sum_{i=1}^{j_l} R_i, \qquad l = 1, 2, \ldots, g,$$

where $j_1, j_2, \ldots, j_g$ are chosen appropriately in advance so that all $E(G_l)$ are likely to be large enough (say > 5). The test statistic is the well known

$$\chi^2_{g-d-1} = \sum_{l=1}^{g} \frac{\{G_l - E[G_l(\hat{\theta})]\}^2}{E[G_l(\hat{\theta})]}.$$

Here $\hat{\theta} = (\hat{\theta}_1, \hat{\theta}_2, \ldots, \hat{\theta}_d)$ is the estimated vector of population parameters and $g - d - 1$ are the degrees of freedom.

When **R** is not multinomially distributed, the problem of testing goodness of fit may be rather intractable. One may still compute the above χ^2-test statistic, provided that it is not used as a test statistic. Under the null hypothesis that the model is correct, the statistic will not have a χ^2-distribution; we can only use it to form a picture of the similarity between $\mathbf{R}$ and $E[\mathbf{R}(\hat{\theta})]$.

Rao (1971) proposed a test based on approximating the distribution of $G_1, G_2, ..., G_g$ by the multivariate normal distribution in the case of a fixed negative binomial model. The test statistic is rather complicated to compute, and no investigation of the validity of the normal approximation was attempted.

Another important problem is that the fixed model, usually containing a huge number of fixed abundances, is extremely restrictive and thus has to be considered as an approximation from the start. We know that the model does not fit from a purely statistical point of view, but it would be wrong to reject its applicability for that reason. A conventional statistical test for goodness of fit therefore seems to be a waste of time. No important practical conclusions can be drawn even from an exact test statistic if it takes non-accepable values. Thus there seems to be a need for other ways of consideration, taking into account that we are willing to accept the model if the 'difference' between model and reality does not seem to be too large. The degree of misfit has to be considered in some sense, and it has to be decided to what extent a misfit can be tolerated. To the author's knowledge no such measures have been proposed in this context.

The following example will be used to illustrate the fact that a poor fit may occur even if there is no noticeable effect on the χ^2-statistic. Let us consider the data of C.B. Williams, i.e. Lepidoptera caught in light traps at Rothamsted in 1934, also dealt with by Bliss (1965) and Bulmer (1974). Table 1 shows the fitted sequence $\{E[G_l(\hat{\theta})]\}_{l=1}^{20}$ for the following cases:

A1: the negative binomial fitted by (pseudo) moments;
A2: the negative binomial fitted by (pseudo) maximum likelihood;
B: the Poisson lognormal model fitted by (pseudo) maximum likelihood;
C: the generalized negative binomial fitted by (pseudo) moments. In our previous notation, the estimation method is $\{(i + 1)^{-1}, 1, i, i^2\}$.

The table also gives estimates of the various population parameters

Table 1. Models fitted to data of C.B. Williams. Lepidoptera caught in a light trap at Rothamsted in 1943; also fitted by Bliss (1965) and Bulmer (1974) $N = 3382$, $S = 176$.

Groups $i +$	Obs $R_i +$	A1 $E(R_i) +$	A2 $E(R_i) +$	B $E(R_i) +$	C $E(R_i) +$
1 +	34	37.63	37.61	31.16	35.39
2 +	19	18.99	18.98	20.81	19.63
3 +	15	12.64	12.64	15.05	13.12
4 +	10	9.43	9.43	11.50	9.74
5 +	10	7.49	7.49	9.14	7.71
6 +	6	6.19	6.19	7.48	6.35
7 +	3	5.26	5.26	6.26	5.38
8 +	14	8.58	8.58	9.92	8.75
10 +	3	6.80	6.80	7.56	6.91
12 +	6	5.59	5.59	5.98	5.66
14 +	12	6.82	6.82	6.97	6.88
17 +	3	5.49	5.49	5.34	5.53
20 +	9	5.90	5.90	5.46	5.93
24 +	3	5.80	5.80	5.08	5.81
29 +	4	6.16	6.17	5.09	6.16
36 +	7	6.31	6.32	4.91	6.29
46 +	2	6.27	6.27	4.62	6.23
61 +	3	5.11	5.11	3.66	5.07
81 +	6	4.69	4.69	3.45	4.65
116 +	7	4.84	4.84	6.55	4.82
	X^2	22.21	22.21	19.56	21.33
	D.f.	17	17	17	16
	$\hat{\kappa}$	0.022	0.022	—	0.004
	$\hat{\alpha}$	41.916	41.945	—	41.236
	$\hat{\beta}$	—	—	—	25 107
	$\hat{\sigma}$	—	—	3.562	—
	$\hat{\mu}$	—	—	− 6.958	—
	$\hat{s}$	—	—	226	262

A1: Negative binomial series fitted by (pseudo) moments.
A2: Negative binomial series fitted by (pseudo) maximum likelihood.
B: Poisson long-normal series fitted by (pseudo) maximum likelihood.
C: Generalized negative binomial series fitted by (pseudo) moments.

and the χ^2-statistics. Further, the estimate of s is given in cases where the model allows that estimation, that is, for B and C. From the point of view of the χ^2-statistic, all models seem to fit well, the lognormal showing the closest fit. Let us now see what would happen if the sample was much smaller. Table 2 shows 10 subsamples of size 100 drawn at random with their corresponding estimated parameters.* The generalized negative binomial has not been fitted since the parameter η (see Section 3.9.) would be virtually zero, and therefore indistinguishable from the negative binomial at this sample size.

Table 2. Subsamples of 100 individuals (labels) drawn at random from the population (sample) given in Table 1

Sample number :	1	2	3	4	5	6	7	8	9	10
R_1	30	31	40	39	19	37	31	31	34	39
R_2	12	12	10	6	5	9	4	10	7	8
R_3	2	2	4	2	7	3	5	4	4	3
R_4	4	2	5	3	4	1	3	3	1	1
R_5	1	2	0	0	2	4	2	1	2	2
R_6	2	1	0	1	0	2	1	1	2	1
R_7	1	1	1	2	0	0	0	2	2	1
R_8	0	1	0	1	3	0	1	0	0	0
R_9	0	0	0	0	0	0	0	0	0	1
R_{10}	0	0	0	0	0	0	1	0	0	0
$R_i, i > 10$	0	0	0	0	0	0	0	0	0	0
κ by A1	0.06	-0.14	-0.07	-0.53	0.47	-0.31	-0.36	-0.09	-0.37	-0.48
α by A1	47.17	34.71	58.47	15.26	43.84	30.56	17.71	38.27	21.55	18.83
κ by A2	0.33	0.01	0.14	-0.30	0.84	0.03	-0.09	0.18	-0.05	-0.33
α by A2	64.33	44.33	76.24	29.75	59.65	54.95	31.26	55.44	40.71	29.17
σ^2 by B	1.15	1.59	1.20	4.49	0.55	1.82	2.86	1.35	2.65	1.12
μ_0 by B	-5.37	-5.80	-8.63	-4.61	-4.70	-6.34	-6.72	-5.57	-6.75	-5.66

Student's t-statistic for testing deviations from the estimates obtained from the complete sample of 3382 individuals

Student's t (9 d.f)

κ by A1 :	0.34
α by A1 :	1.31
κ by A2 :	-2.18
α by A2 :	-1.91
σ^2 by B :	-4.56
μ_0 by B :	2.44

The hypotheses are rejected at the 5% level if $|t| > 2.26$

Student's t-statistic for testing deviations from estimates obtained fitting the generalized negative binomial

κ by A1 :	0.68
α by A1 :	1.44
κ by A2 :	-0.54
α by A2 :	-0.49

*I owe thanks to my wife Aase Fjørtaft for her help with this work.

The example shows clearly that the χ^2-statistic alone may be a little deceptive. The estimates of σ^2 and μ_0 for the lognormal model change significantly (by the conventional Student's t-test) as the sample size is reduced to 100, while this is not the case for the negative binomial. This occurs in spite of the fact that the lognormal gave a smaller χ^2-value. In fact, the lognormal model is not robust against changes in N for this population.

When a sequence of samples is available, one may examine the data with a view to finding systematic misfits. For example, $\Delta = R_1 - E[R_1(\hat{\theta})]$ may tend to have the same sign for all samples. Engen (1975) examined eight different types of samples (birds, insects, marine biological), and it was found that Δ was positive for seven samples when the lognormal was fitted and $\Delta < 0$ for seven samples for the negative binominal. This indicates, expressed in rather inexact terms, that the lognormal gives too small a number of species with small abundances, while the negative binomial gives too many. Fitting the generalized negative binomial is one suggestion towards solving this problem.

To get a further indication of the goodness of fit of the log-normal model we may consider some of the largest order statistics, that is, the abundances of the most abundant species. Now the effect of sampling (the Poisson effect, the Poisson truncation) is negligible for species with large abundances, so we may consider the lognormal instead of the Poisson lognormal distribution. The log-transformed observations can therefore be compared directly with tabulated values for the largest order statistics for the normal distribution. Such tables are given by Harter (1961) and Tippett (1925). For the eight samples mentioned above I have considered the largest order statistics, that is, the most abundant species. Let $O^{(1)}$ be its standardized value (the one relating to the standard normal distribution) and write $E[O^{(1)}]$ for the corresponding tabulated value. The tendency was again quite systematic; $O^{(1)}$ was less than $E[O^{(1)}]$ for all sets of data except one, for which they turned out to be approximately equal. This indicates that the lognormal distribution has too long a tail to fit these data.

The present section does not claim to solve the problem of testing goodness of fit. The purpose of the above remarks was to highlight the need for further research in this area. One ought not in general to be satisfied with an evaluation of the χ^2-statistic only, but ought also to require that the estimators should be robust in withstanding changes in the sample size.

7.3. Quadrat sampling

Introduction

In botanical ecology one is faced with the problem of deciding or defining what is an individual plant. Hence, as mentioned in Section 2.4., the data often tends to be rather incomplete. In practice one frequently places, more or less at random, a number of sampling units (quadrats) in the area to be investigated. For each unit it is possible to obtain a list of the species present, but there are no individual counts.

Write, as in Section 2.4., M_r for the number of species represented in exactly r out of Q quadrats, $r = 0, 1, \ldots, Q$. Note that M_0 is not observable, which is analogous to the fact that R_0 is not observable when the individual counts are complete. We showed in Section 2.4. that $E(M_r)$ is expressible by $g^*(\lambda)$. However, in practice equation (2.14) is usually not appropriate for evaluating $E(M_r)$ when $g^*(\lambda)$ is known. If the sequence $\{E(R_j)\}_{j=1}^{\infty}$ for the total sample (all Q quadrats) is known, the following computational method seems appropriate: suppose as a prior condition that $\mathbf{R}$ is given and assume that any individual is equally likely to occur in any quadrat when it is known to be present in one. Consider one species represented by j individuals in all Q quadrats together. Write $\pi_r(j)$ for the probability that this species is represented in exactly r quadrats ($\pi_{Q-r}(j)$, $r = 1, 2, \ldots, Q$ is sometimes called the classical occupancy distribution, see for example Feller, Vol. I, 1967, p. 60). Then

$$E(M_r|\mathbf{R}) = \sum_{j=1}^{\infty} \pi_r(j)R_j, \quad r = 1, 2, \ldots, Q,$$

and

$$E(M_r) = \sum_{j=1}^{\infty} \pi_r(j)E(R_j). \tag{7.2.}$$

Arguing conditionally on the result for the first $(j-1)$ individuals, we obtain the recurrence relation

$$\pi_r(j) = \pi_r(j-1)\frac{r}{Q} + \pi_{r-1}(j-1)\frac{Q-r+1}{Q}, \tag{7.3}$$

with initial conditions

$$\pi_1(1) = 1, \qquad \pi_r(1) = 0 \quad \text{for } r \neq 1.$$

The summations in (7.2) are now quite simply carried out by using a computer. The sums for $E(M_1), E(M_2), \ldots, E(M_{Q-1})$ usually converge rather quickly since Q is in practice not likely to be very large. $E(M_Q)$ is most conveniently found by subtraction:

$$E(M_Q) = E(S) - \sum_{j=1}^{Q-1} E(M_j).$$

Williams (1950) fitted the sequence $\{E(M_j)\}_{j=1}^Q$ corresponding to Fisher's logarithmic series model to botanical data $M_1, M_2, \ldots, M_Q$. The computations were based on expanding (2.14) binomially and integrating term by term. However, the result is a sum of quite large subsequent positive and negative terms, and one therefore gets trouble with rounding off errors if Q is large. In fact, Williams was not able to evaluate the sequence for $Q > 2$. The method based on (7.2) and (7.3) seems to work well for quite large values of Q.

Estimation

In the case $(R_1, R_2, \ldots)$ conditionally upon $S = \sum_{j=1}^{\infty} R_j$ possesses the multinomial distribution, maximum likelihood estimation may be carried out relatively simply because of the following theorem.

Theorem 7.1. If $(R_1, R_2, \ldots)$ conditionally upon $S = \sum_{j=1}^{\infty} R_j$ is multinomially distributed with parameters $(S, q_1, q_2, \ldots)$, then $M_1, M_2, \ldots, M_Q$ is multinomially distributed with parameters $(S, r_1, r_2, \ldots, r_Q)$ where

$$r_i = \sum_{j=1}^{\infty} \pi_i(j) q_j. \qquad (7.4)$$

Proof. Let $\mathbf{X}, \mathbf{R}$ and $\mathbf{M}$ be defined as before. Provided that individuals, independently of each other, are equally likely to occur in any quadrat, the conditional distribution of $\mathbf{M}$ given $R_1, R_2, \ldots$ depends on $R_1, R_2, \ldots$ and Q only. Hence, the unconditional distribution of $\mathbf{M}$ is uniquely given by the distribution of $\mathbf{R}$.

Assume, without loss of generality, that $X_1, X_2, \ldots, X_s > 0$ and $X_i = 0$ for $i > S$. Suppose that $Pr(X_i = j | X_i > 0) = q_j$ and that the X_i are independent. Then $(R_1, R_2, \ldots)$ is multinomially distributed with parameters $(S, q_1, q_2, \ldots)$ and the distribution of $(M_1, M_2, \ldots$

M_Q) is multinomial $(S, r_1, r_2, \dots, r_Q)$, where $r_i = \Sigma_i \pi_i(j) q_j$. The latter statement follows directly from the way $\pi_i(j)$ is defined. By the first part of this proof the distribution of $(M_1, M_2, \dots, M_Q)$ is uniquely given by the distribution of $(R_1, R_2, \dots)$. Therefore the underlying assumption concerning the distribution of $\mathbf{X}$ is irrelevant as long as $(R_1, R_2, \dots)$ is a multinomial variate. This completes the proof.

For any model where $(R_1, R_2, \dots)$ is multinomially distributed, the likelihood function is

$$L = \sum_{i=1}^{Q} M_i \ln r_i = \sum_{i=1}^{Q} M_i \ln \left[\sum_{j=1}^{\infty} \pi_i(j) q_j \right].$$

Further, if θ is a population parameter,

$$\frac{\partial L}{\partial \theta} = \sum_{i=1}^{Q} M_i \frac{1}{r_i} \frac{\partial r_i}{\partial \theta},$$

where $\dfrac{\partial r_i}{\partial \theta}$ is often most conveniently evaluated by

$$\frac{\partial r_i}{\partial \theta} = \sum_{j=1}^{\infty} \pi_i(j) \frac{\partial q_i}{\partial \theta}, \qquad i = 1, 2, \dots, Q - 1,$$

and

$$\frac{\partial r_Q}{\partial \theta} = - \sum_{i=1}^{Q-1} \frac{\partial r_i}{\partial \theta},$$

which is just the derivative of (7.3).

If $(R_1, R_2, \dots)$ is not a multinomial variable, this estimation may still be carried out, though it is no longer the maximum likelihood method ('pseudo maximum likelihood').

Analogous to our treatment of the negative binomial model (see Section 3.4 pp. 75) we may alternatively evaluate the 'moments', such as $\Sigma i E(M_i)$, $\Sigma i^2 E(M_i)$. I have worked out a computer programme to carry out estimation for the parameters α and κ in the negative binomial model for the case of incomplete sampling (Engen, 1976, unpublished).

The programme also evaluates first-order approximations to the standard errors for the fixed model. (These are the results of quite tedious but uncomplicated algebra which cannot be reproduced here).

An extension of the general assumptions

Before we proceed to cite an example, it is worth pointing out that we can extend our general assumptions somewhat without altering the distribution of M. What has so far been interpreted as individuals may be viewed as groups (flocks, patches) of individuals. Members of one group are all so close to each other that they all occur in the same quadrat if one of them does. Hence, X_i is now the number of groups of species C_i in all Q quadrats together. Now the intepretation of M_i is unaltered, but it is necessary to investigate under which assumptions $E(X_i)/N = p_i$ can still be regarded as the relative abundance of species C_i. Let $Z_1, Z_2, \ldots, Z_{X_i}$ denote the number of individuals in the various groups of species C_i occurring in the sample and write

$$Y_i = Z_1 + Z_2 + \ldots + Z_{X_i}$$

for the corresponding number of individuals caught. Suppose the Z_j are independently and identically distributed with

$$E(e^{tZ_i}) = M_Z(t).$$

Hence

$$M_{Y_i}(t) = G_{X_i}[M_Z(t)]. \tag{7.5}$$

By (7.5) it is straightforward to check that Y_i/N converges to $p_i E(Z)$ in probability as $N = \Sigma X_i \to \infty$ if var $(Z) < \infty$. Therefore, *if the expected group size is the same for all species,* then $\Sigma Y_i/N$ converges to $E(Z)$ in probability, and $Y_i/\Sigma Y_j$ converges to p_i. Hence $E(X_i)/N = p_i$ is actually the relative abundance of species C_i, and the interpretation of the parameters of the distribution of M is unaltered by our generalization.

Christmas counts is Sweden

S. Svensson has kindly let me see some of his data from so-called 'Christmas counts' of bird species in Sweden. The counter moves from one site to another, stops for a prescribed period of time and writes down the species he observes. The data in Table 3 are compiled by one counter who stopped at 20 sites and recorded the species observed at each site within a period of 5 minutes. Professor S. Haftorn has informed me that these species have a tendency to occur in flocks. The mean size of a flock would probably be of order 5, but the number of individuals varies from one flock to another.

Table 3. Svensson's Christmas counts. The negative binomial model fitted to incomplete data

	Data	By pseudo maximum likelihood method	By pseudo moment method
i	M_i	$E(M_i)$	$E(M_i)$
1	14	14.41	14.95
2	6	6.03	5.90
3	5	3.56	3.41
4	3	2.39	2.28
5	1	1.73	1.64
6	0	1.30	1.24
7	2	1.00	0.96
8	0	0.79	0.76
9	0	0.53	0.61
10	1	0.50	0.50
11	0	0.40	0.40
12	1	0.32	0.33
13	0	0.26	0.27
14	0	0.20	0.22
15	1	0.16	0.17
16	0	0.12	0.13
17	0	0.09	0.10
18	0	0.06	0.07
19	0	0.04	0.05
20	0	0.11	0.07
Total	34	34.0	34.0
estimate of α		11.707 ± 1.387	10.043 ± 4.059
estimate of κ		-0.107 ± 0.258	-0.167 ± 0.163
estimated correlation		0.7230	0.9773
estimate of v^*		6.75	6.85
estimate of Qv^*		135	137
information index		3.270 ± 0.427	3.248 ± 0.085
Simpson's index		0.930 ± 0.015	0.925 ± 0.014
estimate of $E(Z)$		7.70	7.75

A flock may also consist of one single individual. Table 3 shows the data and the fitted sequence $[E(M_i)]_{i=1}^Q$ derived from the negative binomial. The model is fitted by (pseudo) maximum likelihood and (pseudo) 'moments'. Standard deviations refer to the fixed model, and v^* denotes the expected number of flocks per 'quadrat'. The counter mentioned above also tried to count (approximately) the number of individuals, which he estimated at 1024. Hence the expected flock size, $E(Z)$, is estimated by $1024/(Qv^*)$, which is about 8. I have also fitted the model to three other sets of data giving $E(Z)$ varying from 4 to about 8, which agrees well with what the experts know about flock sizes for these bird species.

It seems difficult to recommend any particular estimation method in general. For example, in Table 3 the pseudo maximum likelihood method estimates α with higher precision than does the pseudo moment method, while the later method is preferable with respect to κ. The pseudo moment method is superior with respect to calculating time and expenses. For large samples the standard errors will be small in any case, and there is therefore little point in adopting the likelihood theory in this situation. For smaller samples, it seems preferable to evaluate standard errors for both methods before making the choice.

7.4. Higher order classification

C.B. Williams demonstrated in several papers different applications of the logarithmic series to ecological as well as to other problems. In one paper (Williams, 1944), he also considered the grouping of species into genera. This is analogous to what we have dealt with so far, except that we are now faced with a higher order of classification. Instead of grouping individuals into species, the grouping of species into genera is considered. It must be emphasized that our theory of sampling is not valid for this application, since it is definitely not correct to claim that we are selecting species at random so that each species has the same probability of being chosen. In fact, this is not a sampling situation at all; a known set of species is just classified into genera according to certain principles. The random element lies in evolution and in the act of classifying. But it is the underlying continuous ('approximately continuous') 'genera abundance distribution' and not the Poisson assumption that is crucial in deciding the pattern of the observations $R_1, R_2, \dots$, and it is certainly one step towards a better understanding of evolution to arrive at mathematical models that fit well to species genera data. In order to be realistic, any qualitative evolutionary theory must be in accordance with the observed patterns.

One cannot claim that the application of the logarithmic series to this problem was successful. Though the tail of the series always fitted reasonably well, there seemed to be an overall problem with the first terms, especially the term $E(R_1)$ which in most cases was definitely underestimated by the model. Williams noticed this misfit, writing (the following quotation also deals with his species-individuals data):

There is, however, one general inconsistency that must be noted. In the present paper calculated and observed series are given for 22 sets of data. In only one of them is the observed R_1 definitely below the calculated; in two more it is slightly below; in two almost identical; and in 17 cases the observed R_1 is above the calculated. In the paper on light-trap captures in Harpenden already published (Fisher *et al.*, 1943), there were 27 sets of figures, in 19 of which the observed R_1 was above the calculated and in only 8 below. In some yet unpublished analyses of light-trap captures of Lepidoptera in U.S.A. I have calculated 11 series of frequency distributions and they all gave calculated R_1 below the observed; sometimes the difference is small, but in others quite large; in one case the calculated R_1 is 10 and the observed 19. There seems therefore to be a general tendency for the logarithmic series to underestimate the number of groups with one individual, as compared to the observed facts. Either the theory is incomplete or there is some biased error in the observations producing too many very small groups.

I think Williams' first suggestion is likely to be true. The logarithmic distribution, having only one parameter, is a very simple distribution, and one does not really have any reason at all to believe in its applicability in cases where misfits as systematic as in these examples are found.

The difference between the observed and estimated R_1 is most significant for the species genera data. Table 4 shows the extended negative binomial fitted by (pseudo) maximum likelihood to some of the data considered by Williams. They also show clearly the vast improvement obtained for the first terms when we move from the logarithmic to the extended negative binomial model.

7.5. Species diversity of chironomid communities—an example

The following example taken from Aagaard and Engen (1977) illustrates to what extent 'biologically' and 'physically' controlled (Odum, 1971, p. 148) chironomid communities are separated in an (α, κ) diagram when the negative binomial series is fitted. In Figure 2 each closed curve represents a sample, the centre being the estimate

Table 4. Frequencies of genera with different number of species in British insects
British macrolepidoptera (excluding butterflies)

Number of species per genus	From Stainton (1857)			From South and Edelstein (1939)		
i	obs. R_i	log. $E(R_i)$	ext. n. b. $E(R_i)$	obs. R_i	log. $E(R_i)$	ext. n. b. $E(R_i)$
1	132	112.13	132.38	239	203.40	240.10
2	40	47.12	42.33	51	72.68	49.80
3	25	26.40	21.81	25	34.62	22.19
4	12	16.64	13.37	12	18.56	12.53
5	13	11.19	8.98	8	10.61	7.97
6	8	7.84	6.38	3	6.32	5.45
7	3	5.64	4.71	4	3.87	3.91
8	3	4.15	3.58	3	2.42	2.91
9	3	3.10	2.78	2	1.54	2.22
10	0	2.35	2.19	3	0.99	1.73
11	3	1.79	1.75	1	0.64	1.38
12	0	1.38	1.42	0	0.42	1.10
13	1	1.07	1.16	1	0.28	0.89
14	0	0.84	0.95	0	0.18	0.73
15	2	0.66	0.79	1	0.12	0.60
16	0	0.52	—	0	0.08	0.50
17	0	0.41	—	2	0.06	0.42
18	1	0.32	—	1	0.04	0.35
19	0	0.26	—	0	0.03	—
20	0	0.21	—	0	0.02	—

Also at 23, 26 and 40		Also at 43	
For log. model $\alpha = 133.4$		For log. model $\alpha = 284.59$	
For ext. n. b. model $\kappa = -0.294$ $\alpha = 72.8$		For ext. n. b. model $\kappa = -0.550$ $\alpha = 60.5$	

obs.   = observed series
log.   = logarithmic series
ext. n. b. = extended negative binomial series

$(\hat{\alpha}, \hat{\kappa})$ found by the (pseudo) moment method. The curves indicate approximations to standard errors 'in any direction'. These are arrived at by considering some rotation, say an angle θ, of the diagram, giving

$$\hat{\alpha}' = \hat{\alpha}\cos\theta + \hat{\kappa}\sin\theta$$

$$\hat{\kappa}' = -\hat{\alpha}\sin\theta + \kappa\cos\theta.$$

SD $(\hat{\alpha}')$ and SD $(\hat{\kappa}')$ may be evaluated as functions of var $(\hat{\alpha})$, var $(\hat{\kappa})$ and $\rho(\hat{\alpha}, \hat{\kappa})$, which are tabulated by Engen (1974). Further these standard deviations are plotted in the appropriate direction.

The data used are based on emergence trap collections published by Sandberg (1969) and Laville (1971). Sandberg's data (white areas) are taken from the littoral zone (1–4m) of Lake Erken in Sweden. The diversity of the community in this moderately entrophic lake, is reckoned to be 'biologically controlled'. The black areas represent data from lakes in the French Haute-Pyrénées, the littoral and profundal zones (1–19m) (Laville, 1971). These lakes are oligotrophic, and there are reasons to believe that the diversity is controlled by physical factors. Fig. 3 shows how these communities are quite clearly separated in the (α, κ) diagram.

In Fig. 2, values of H_s, H and H_v (see Chapter 5) are indicated in the same diagram. A comparison of the two figures reveals that H_s and H_v are incapable of separating the blacks from the whites, while H, at least for these data, would perform this task quite well on its own. It is a challenge to research workers in ecology to explore the factors responsible for the variation, for example, in H_v and α.

Sandberg (1969) used the broken stick model and found a poor fit. This is in agreement with Fig. 3, all values of κ being far from 1 (see Chapter 3). In fact, the logarithmic series ($\kappa = 0$) would work better, though it is not quite satisfactory. (Note that standard errors tend to be larger as α and κ increase; this is an effect of scaling and we should not pay too much attention to it.)

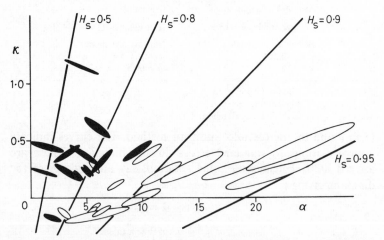

Fig. 3 *The negative binomial model fitted to data from Chironomid communities.*
White areas: biologically controlled diversity.
Black areas: physically controlled diversity.
(Redrawn from Aagaard and Engen, 1977)

References

Aagaard, K. and Engen, S. (1977) Species diversity of chironomid communities, *Acta Univ. Caralinae* (in press).

Abramowitz, M. and Stegun, I.A. (1964) *Handbook of Mathematical Functions with Formulas, Graphs and Mathematical Tables.* National Bureau of Standards, Applied mathematics series SS, U.S. Government Printing office, Washington D.C.

Aitchison, J. and Brown, J.A.C. (1969) *The Lognormal Distribution with Special References to its uses in Economics*, Cambridge University Press.

Anscombe, F.J. (1950) Sampling theory of the negative binomial and the logarithmic series distributions, *Biometrika* **37**, 651–60.

Arrhenius, O. (1921) Species and area, *J. Ecol.* **9**, 95–99.

Basharin, G.P. (1959) On statistical estimate for the entropy of a sequence of independent random-variables, N. Artin (editor), *Theory of Probability and its Applications*, Vol. IV, (Translation of Teoriya Veroyatnostei i ee primeneniya) Society for Industrial and Applied Mathematics, Philadelphia, 333–336.

Bliss, C.I. (1965) An analysis of some insect trap records, *Classical and Contagious Discrete Distributions*, G.P. Patil (editor), Pergamon Oxford, 385–97.

Boswell, M.T. and Patil, G.P. (1971) Change mechanisms generating the logarithmic series distribution used in the analysis of number of species and number of individuals, *Statistical Ecology*, Vol. I, G.P. Patil, E.C. Pielou and W.E. Waters (editors), Pennsylvania State University Press, 99–130.

Bowman, K.O. and Shenton, L.R. (1969) Properties of the maximum likelihood estimator for the parameter in the logarithmic series distribution, *Random Counts in Scientific work*, G.P. Patil (editor.), The Pennysylvania State University Press, University Park.

Brillouin, L. (1962) *Science and Information Theory* 2nd ed. Academic Press, New York.

Brown, S. and Holgate, P. (1971) Tables of the Poisson lognormal distribution, *Indian J. Statistics* **33**, Series B, 235–58.

Bulmer, M.G. (1974) On fitting the Poisson lognormal distribution to species-abundance data. *Biometrics* **30**, 101–10.

Buzas, M.A. and Gibson, T.G. (1969) Species diversity: benthonic Foramini-fera in western North Atlantic, *Science* **163**, 72–75.

Cohen, J.E. (1966) *A Model of Simple Competition*, Harvard University Press.

Cohen, J.E. (1968) Alternative derivation of a sepcies abundance relation, *Am. Nat.* **102**, 165–71.

Corbet, A.S. (1942) The distribution of butterflies in the Malay Peninsula. *Proc. R. Ent. Soc. Lond.* (A) **16**, 101–16.

Dirk, C.O. (1938) *Biological Studies of Maine Moths by Light Trap Methods*, Bulletin 389. Maine Agricultural Experimental Station.

Eberhardt, L.L. (1969) Some aspects of species diversity models, *Ecology* **50**, 503–5.

Eldridge, R.C. (1911) *Six Thousand Common English Words*, Privately printed at Niagara Falls, New York.

Engen, S. (1974) On species frequency models, *Biometrika* **61**, 263–70.

Engen, S. (1975a) A note on the geometric series as a species frequency model, *Biometrika* **62**, 697–99.

Engen, S. (1975b) *Statistical Analysis of Species Diversity*, D. Phil. thesis, University of Oxford.

Engen, S. (1975c) The coverage of a random sample from a biological com-munity, *Biometrics* **31**, 201–8.

Engen, S. (1976a) A note on the estimation of the species-area curve, *J. Cons. int. Explor. Mer* **36**, 286–88.

Engen, S. (1976b) *The Distribution of Species on Quadrats*, unpublished manuscript.

Engen, S. (1977a) Exponential and logarithmic spcies – area curves, *Am. nat.* **111**, 591–94.

Engen, S. (1977b) Comments on two different approaches to the analysis of species frequency data, *Biometrics* **33**, 205–213.

Feller, W. (1968) *An Introduction to Probability Theory and its Applications*, Vol. I. John Wiley & Sons, Inc. New York.

Fisher, R.A., Corbet, A.S. and Williams, C.B. (1943) The relation between the number of species and the number of individuals in a random sample from an animal population. *J. Anim. Ecol.* **12**, 42–58.

Fisher, R.A. (1956) *Statistical Methods and Scientific Inference*, Oliver and Boyd, Edinburgh.

Fraser, D.A.S. (1968) *The Structure of Inference*, John Wiley & Sons, Inc. New York.

Gleason, H.A. (1922) On the relation between species and area, *Ecology* **3**, 156–62.

Good, I.J. (1953) The population frequencies of species and the estimation of population parameters, *Biometrika* **40**, 237–64.

Goodall, D.W. (1952) Quantitative aspects of plant distribution, *Biol. Rev.* **27**, 194–245.

Goodman, L.A. (1949) On the estimation of the number of classes in a population, *Ann. Math. Statist.* **20**, 572–9.

Grundy, R.M. (1951) The expected frequencies in a sample of an animal population in which the abundancies are lognormally distributed, *Biometrika* **38**, 427–34.

Harter, H.L. (1961) Expected values of the normal orderstatistics, *Biometrika* **48**, 151–65.

Holgate, P. (1969) Species frequncy distributions, *Biometrika* **56**, 651–60.

Holthe, T. (1975) A method for the calculation of ordinate values of the cumulative species – area curve, *J. Cons. int. Explor. Mer* **36**, 183–4.

Hutcheson, K. (1970) A test for comparing diversities based on the Shannon formula, *J. Theor. Biol.* **29**, 151–54.

Hutcheson, K. and Shenton, L.R. (1974) Some moments of an estimate of Shannon's measure of information, *Communication in Statistics* **3**, 89–94.

Jaccard, P. (1902) Lois de distribution florale dans la zone alpine, *Bull. Soc. vaud. Sci. nat.* **38**, 69–130.

Johnson, N.L. and Brinkhurst, R.O. (1971) Associations and species diversity in benthic macroinvertebrates of Bay of Quinte and Lake Ontario, *J. Fish. Res. Bd. Can.* **28**, 1683–97.

Karlin, S. and Mcgregor, J. (1967) The number of mutant forms maintained in a population, *Proc. 5th Berkeley Symp. Math. Statist. Prob.* **IV**, 415–438.

Kempton, R.A. (1975) A generalized form of Fisher's logarithmic series, *Biometrika* **62**, 29–38.

Kendall, D.G. (1948) On some modes of population growth leading to R.A. Fisher's logarithmic series distribution, *Biometrika* **35**, 6–15.

Kendall, M.G. and Stuart, A. (1969) *The Advanced Theory of Statistics*, Vol. 1, 3rd ed. Griffin, London.

Kilburn, P. (1963) Exponential values for the species – area relation, *Science* **141**, 1276.

Kilburn, P. (1966) Analysis of the species – area relation, *Ecology* **47**, 831–43.

Krylow, V.V. (1971) On the station – species curve, *Statistical Ecology*, Vol. III, G.P. Patil, E.C. Pielou and W.E. Waters (editors), Pennsylvania State University Press, 233–235.

Laville, H. (1971) Recherches sur les chironomides (Diptera) lacustres du Mass if de Néouvielle (Hautes-Pyrénées), *Annls Limnol.* **7**, 1973–414.

Lloyd, M. and Ghelardi, R.J. (1964) A table for calculating the equitability component of species diversity, *J. Anim. Ecol.* **33**, 217–25.

Longuet-Higgins, M.S. (1971) On the Shannon – Wiever index of diversity in relation to the distribution of species in bird censuses. *Theor. Pop. Biol* **2**, 271–89.

MacArthur, R.H. (1957) On the relative abundance of bird species, *Proc. Nat. Acad. Sci. Washington*, **43**, 293–95.

MacArthur, R.H. (1960) On the relative abundance of species, *Am. Nat.* **94**, 25–36.

MacArthur, R.H. (1964) Environmental factors affecting bird species diversity, *Am. Nat.* **98**, 387–97.

MacArthur, R.H. (1965) Patterns of species diversity, *Biol. Rev.* **40**, 510–33.

MacArthur, R.H., Recher, H. and Cody, M. (1966) On the relation between habitat selection and species diversity, *Amer. Nat.* **100**, 319–32.

Maritz, J.S. (1970) *Empirical Bayes Methods*, Methuen, London.

Mehninick, E.F. (1964) A comparison of some species – individuals diversity indices applied to samples of field insects, *Ecology* **45**, 859–61.

Odum, E.P. (1971) *Fundamentals of Ecology*, 3rd ed. W.B. Saunders Company. Philadelphia, London, Toronto.

Patil, G.P. (1962) Some methods of estimation for the logarithmic series distribution, *Biometrics* **18**, 68–75.

Patil, G.P. and Bildiker, S. (1966) On minimum variance unbiased estimation for the logarithmic series distribution, *Sankhyā Ser.* **A 28**, 239–250.

Patil, G.P. and Joshi, S.W. (1968) *A Dictionary and Bibliography of Discrete Distributions*, Oliver and Boyd, Edinburgh.

Patil, G.P. and Wani, J.K. (1965) Maximum likelihood estimation for the complete and truncated logarithmic series distributions, *"Classical and Contagious Discrete Distributions"*, G.P. Patil (editor), Statistical Publishing Co., Calcutta.

Pielou, E.C. (1966) Species diversity and pattern diversity in the study of ecological succession. *J. Theor. Biol.* **10**, 370–83.

Pielou, E.C. (1969) *An Introduction to Mathematical Ecology*, Wiley-Interscience, New York.

Pielou, E.C. (1975) *Ecological diversity*, Wiley, New York.

Preston, F.W. (1948) The commonness and rarity of species, *Ecology* **29**, 254–83.

Quenouille, M.H. (1956) Notes on bias in estimation, *Biometrika* **43**, 353–60.

Rao, C.R. (1971) Some comments on the logarithmic series distribution in the analysis of insect trap data, *Statistical Ecology*, Vol. I, G.P. Patil E.C. Pielou and W.E. Waters (editors), Pennsylvania State University Press, 131–42.

Renyi, A. (1961) On measures of entropy and information, *Proc. 4th. Berkeley Symposium on Math. and Prob.* **1**, 547–561.

Ross, G.J.S. (1970) The efficient use of function minimization in non-linear maximum likelihood estimation, *J.R. Statist. Soc.* **C 19**, 205–21.

Sampford, M.R. (1955) The truncated negative binomial distribution, *Biometrika* **42**, 58–69.

Sandberg, G. (1969) A quantitative study of chironomid distribution and emergence in Lake Erken. *Arch. Hydrobiol. Suppl.* **35**, 119–201.

Saunders, A.A. (1936) *Ecology of the Birds of Quaker Run Valley, Alleghany State Park, New York*. Handbook N.Y. St. Mus. 16.

Shannon, C.E. and Weaver, W. (1949) *The Mathematical Theory of Communication*, University of Illinois Press, Urbana.

Sheldon, A.L. (1969) Equitability indices: dependence upon species count, *Ecology* **50**, 466–67.

Simpson, E.H. (1949) Measurement of diversity, *Nature* **163**, 688.

South, R. and Edelsten, H.M. (1939) *The Moths of the British Isles*, New edition.

Stainton, H.T. (1857) *A manual of British Butterflies and Moths*, London.

Taylor, L.R. and Kempton, R.A. (1974) Log-series and log-normal parameters as discriminants for the Lepidoptera, *J. Anim. Ecol.* **43**, 381–99.

Tippett, L.H.C. (1925) On the extreme individuals and the range of samples taken from a normal population, *Biometrika* **17**, 364.

Tricomi, F.G. (1955) *Vorelesungen Über Orthogonalreihen*, Springer-Verlag, Berlin.

Watterson, G.A. (1974) Models for the logarithmic species abundance distributions, *Theor. Pop. Biol.* **6**, 217–50.

Webb, D.J. (1974) The statistics of relative abundance and diversity. *J. Theor. Biol.* **43**, 277–91.

Wette, R. (1959) Zur biomathematischen Begründung der Verteilung der Elemente taxomischer Einheiten des natürlichen Systems in einer logarithmischen Reihe, *Biom. Zeit.* **1**, 44–50.

Whittaker, R.H. (1972) Evolution and measurements of species diversity. *Taxon* **21**, 213–51.

Williams, C.B. (1944) Some applications of the logarithmic series and the index of diversity to ecological problems, *J. Ecol.* **32**, 1–44.

Williams, C.B. (1946) The logarithmic series and the comparison of island floras, *Proc. Linn. Soc. London.* **159**, 104–8.

Williams, C.B. (1947) The logarithmic series and its application to biological problems, *J. Ecol.* **34**, 253–72.

Williams, C.B. (1950) Application of the logarithmic series to the frequency of occurrence of plant species in quadrats, *J. Ecol.* **38**, 107–38.

Williams, C.B. (1964) *Patterns in the balance of nature*, Academic Press, London.

Williamson, E. and Bretherton, M.H. (1964) Tables of the logarithmic series distribution, *Ann. Math. St.* **35**, 284–97.

Willis, J.C. (1922) *Age and Area*, University Press, Cambridge.

Yule, G.U. (1944) *The Statistical Study of Literary Vocabulary*, University Press, Cambridge.

Zipf, G.K. (1932) *Selected Studies of the Principle of Relative Frequency in Language*, Harvard University Press.

Appendix A
The structural distribution for the model dealt with in example 2.4

We consider the population with relative abundances $(q_1, q_2, \dots)$ defined by

$$q_1 = Q_1, q_i = Q_i \prod_{j=1}^{i-1} (1 - Q_j), \quad i = 1, 2, \dots,$$

where the Q_i are independent beta variates with parameters $(\kappa + 1, \alpha - i\kappa)$ respectively, and where $\alpha > 0, -1 < \kappa < 0$. Our aim is to find the structural distribution for this model.

If the class with abundance q_1 is excluded from the population, the relative abundances are changed to

$$q_2' = Q_2, q_i' = Q_i \prod_{j=2}^{i-1} (1 - Q_j), \quad i = 3, 4, \dots.$$

Consequently, the distribution of $(q_2', q_3', \dots)$ is the same as that of $(q_1, q_2, \dots)$ with α replaced by $\alpha - \kappa$. Now, write Z for the abundance of the class of an element chosen randomly from the population $(q_1, q_2, \dots)$ and Z' for the corresponding random variable defined for $(q_2', q_3', \dots)$. Write

$$\zeta(\alpha, \kappa, t) = (E(Z^t), t = 0, 1, \dots.$$

Then

$$\zeta(\alpha - \kappa, \kappa, t) = E(Z'^t).$$

By a straightforward conditional argument we obtain

$$\zeta(\alpha, \kappa, t) = E\{q_1^t P(Z = q_1) + (1 - q_1)^t \zeta'^t [1 - P(Z = q_1)]\},$$

leading to the equation

$$\zeta(\alpha, \kappa, t) = E[q_1^{t+1}] + E[(1 - q_1)^{t+1}] \zeta(\alpha - \kappa, \kappa, t). \tag{A1}$$

Inserting

$$E(q_1^t) = \frac{\Gamma(\kappa + 2 + t)\Gamma(\alpha + 1)}{\Gamma(\kappa + 1)\Gamma(\alpha + 2 + t)}$$

and

$$E[(1 - q_1)^t] = \frac{\Gamma(\alpha - \kappa + 1 + t)\Gamma(\alpha + 1)}{\Gamma(\alpha - \kappa)\Gamma(\alpha + 2 + t)},$$

it is straightforward to check that

$$\zeta^*(\alpha, \kappa, t) = \frac{\Gamma(\alpha + 1)\Gamma(\kappa + 1 + t)}{\Gamma(\kappa + 1)\Gamma(\alpha + 1 + t)} \qquad (A2)$$

fits in (A1). It remains to be shown that this is the only solution.

Suppose we had two solutions ψ_1 and ψ_2 and write $\psi_0 = \psi_1 - \psi_2$. Inserting the solutions and subtract the equations we get

$$\zeta_0(\alpha, \kappa, t) = E[(1 - q_1)^t]\,\zeta_0(\alpha - \kappa, \kappa, t). \qquad (A3)$$

(A3) may be written

$$\varphi_0(\alpha, \kappa, t) = \left(\frac{\alpha + 1 + t}{\alpha}\right)\varphi_0(\alpha - \kappa, \kappa, t), \qquad (A4)$$

where

$$\varphi_0(\alpha, \kappa, t) = \zeta_0(\alpha, \kappa, t)\alpha(\alpha + 1)\ldots(\alpha + t).$$

From (A4) we find by induction

$$\varphi_0(\alpha, \kappa, t) = \frac{\alpha(\alpha - \kappa)(\alpha - 2\kappa)\ldots[\alpha - (n - 1)\kappa]}{(\alpha + 1 + t)(\alpha - \kappa + 1 + t)\ldots[a - (n - 1)\kappa + 1 + t]}$$

$$\times \varphi_0(\alpha - n\kappa, \kappa, t).$$

Now the first factor is, asymptotically as $n \to \infty$, of order $n^{(1+t)/\kappa}$. Further $\psi(\alpha, \kappa, t) = E(Z^t)$ is at most of order 1, and $\varphi_0(\alpha - n\kappa, \kappa, t)$ is of order less than n^{t+1}. It follows that

$$0 \leqslant \varphi_0(\alpha, \kappa, t) \leqslant \lim_{n \to \infty} Cn^{(t+1)/\kappa + t + 1} = C \lim n^{(t+1)[(\kappa+1)/(\kappa)]} = 0,$$

$$\text{since } \frac{\kappa + 1}{\kappa} < 0.$$

Hence, the structural distribution is the beta distribution with parameters $(\kappa + 1, \alpha - \kappa)$.

Appendix B

Table 1. $\sqrt{N}\sigma(p)$ as a function of κ and ω

	ω						
κ	0.100	0.090	0.080	0.070	0.060	0.050	0.040
$-.90$	.50473	.51010	.51560	.52123	.52700	.53294	.53903
$-.80$	.60990	.61163	.61286	.61346	.61320	.61176	.60857
$-.70$	.63928	.63624	.63212	.62666	.61945	.60990	.59703
$-.60$	.63278	.62511	.61593	.60488	.59147	.57493	.55406
$-.50$	.60744	.59573	.58224	.56658	.54819	.52627	.49954
$-.40$	.57226	.55727	.54036	.52114	.49906	.47335	.44275
$-.30$	.53244	.51491	.49545	.47368	.44911	.42099	.38821
$-.20$	.49110	.47174	.45050	.42707	.40100	.37163	.33800
$-.10$	.45015	.42955	.40722	.38286	.35610	.32639	.29291
.00	.41091	.38961	.36681	.34185	.31526	.28581	.25296
.10	.37343	.35183	.32886	.30432	.27796	.24943	.21820
.20	.33866	.31712	.29440	.27036	.24481	.21746	.18792
.30	.30652	.28531	.26312	.23984	.21534	.18940	.16175
.40	.27703	.25635	.23489	.21256	.18927	.16489	.13919
.50	.25010	.23011	.20951	.18826	.16629	.14351	.11979
.60	.22561	.20642	.18678	.16668	.14608	.12492	.10313
.70	.20341	.18509	.16647	.14755	.12832	.10876	.08883
.80	.18332	.16592	.14835	.13062	.11274	.09472	.07655
.90	.16518	.14871	.13219	.11564	.09908	.08253	.06601
1.0	.14882	.13329	.11781	.10240	.08710	.07194	.05695
1.2	.12080	.10711	.09361	.08036	.06738	.05473	.04247
1.4	.09809	.08612	.07445	.06314	.05221	.04171	.03172
1.6	.07970	.06930	.05927	.04966	.04050	.03183	.02373
1.8	.06479	.05580	.04723	.03909	.03144	.02431	.01776
2.0	.05270	.04496	.03765	.03080	.02443	.01858	.01329

	ω					
κ	0.030	0.020	0.015	0.010	0.006	0.003	0.001
$-.90$	.54529	.55159	.55463	.55731	.55857	.55742	.55080
$-.80$	.60260	.59152	.58228	.56794	.54846	.52077	.47632
$-.70$	.57905	.55214	.53253	.50481	.47051	.42615	.36246
$-.60$	.52672	.48838	.46187	.42601	.38385	.33241	.26398
$-.50$	.46576	.42036	.39011	.35051	.30576	.25371	.18875

Table 1. (*Contd.*)

κ	ω						
	0.030	0.020	0.015	0.010	0.006	0.003	0.001
− .40	.40515	.35627	.32468	.28447	.24058	.19162	.13393
− .30	.34886	.29917	.26792	.22913	.18809	.14405	.09481
− .20	.29846	.24982	.21999	.18382	.14663	.10812	.06714
− .10	.25429	.20794	.18016	.14719	.11422	.08119	.04764
.00	.21615	.17274	.14738	.11782	.08902	.06107	.03390
.10	.18341	.14345	.12051	.09434	.06945	.04603	.02421
.20	.15554	.11910	.09859	.07563	.05429	.03479	.01736
.30	.13188	.09893	.08073	.06071	.04252	.02637	.01249
40	11185	..08224	.06618	.04882	.03338	.02004	.00902
.50	.09491	.06843	.05432	.03932	.02625	.01527	.00653
.60	.08058	.05700	.04465	.03172	.02069	.01166	.00474
.70	.06847	.04753	.03674	.02562	.01633	.00891	.00345
.80	.05822	.03967	.03027	.02073	.01273	.00683	.00251
.90	.04954	.03315	.02497	.01679	.01022	.00523	.00183
1.0	.04219	.02772	.02061	.01361	.00809	.00401	.00133
1.2	.03066	.01942	.01407	.00895	.00508	.00236	.00070
1.4	.02232	.01363	.00962	.00589	.00319	.00138	.00037
1.6	.01627	.00957	.00658	.00388	.00200	.00081	.00019
1.8	.01186	.00672	.00449	.00255	.00125	.00047	.00010
2.0	.00864	.00471	.00306	.00167	.00078	.00028	.00005

Table 2. $\sigma(-plnp)$ *as a function of* α, κ *and* N

$N = 100$

κ	α						
	2	3	5	7	10	15	20
− .50	.08736	.08621	.08621	.08668	.08729	.08776	.08771
− .40	.08369	.08246	.08302	.08416	.06564	.08725	.08806
− .30	.07840	.07705	.07814	.07988	.08220	.08494	.08664
− .20	.07250	.07096	.07244	.07470	.07776	.08154	.08408
− .10	.06658	.06475	.06648	.06915	.07281	.07749	.08079
.00	.06095	.05874	.06060	.06355	.06767	.07310	.07708

$N = 200$

κ	α						
	2	3	5	7	10	15	20
− .50	.05697	.05619	.05632	.05687	.05770	.05877	.05948
− .40	.05257	.05175	.05223	.05317	.05454	.05636	.05768
− .30	.04751	.04663	.04738	.04865	.05047	.05295	.05481
− .20	.04244	.04145	.04237	.04388	.04606	.04906	.05139
− .10	.03769	.03656	.03756	.03921	.04163	.04502	.04772
.00	.03341	.03209	.03310	.03483	.03739	.04104	.04401

Table 2. (*Contd.*)

$N = 300$

κ	α						
	2	3	5	7	10	15	20
$-.50$	.04414	.04349	.04359	.04406	.04481	.04585	.04663
$-.40$	.03986	.03919	.03954	.04029	.04142	.04301	.04424
$-.30$	.03528	.03457	.03510	.03606	.03750	.03953	.04114
$-.20$	.03088	.03011	.03074	.03185	.03349	.03584	.03775
$-.10$	.02690	.02603	.02670	.02788	.02965	.03221	.03433
.00	.02340	.02242	.02307	.02427	.02609	.02877	.03102

$N = 500$

κ	α						
	2	3	5	7	10	15	20
$-.50$	.03188	.03137	.03141	.03176	.03234	.03320	.03390
$-.40$	.02802	.02750	.02771	.02823	.02906	.03027	.03126
$-.30$	.02416	.02362	.02393	.02458	.02558	.02705	.02827
$-.20$	.02062	.02005	.02042	.02114	.02224	.02387	.07524
$-.10$	.01753	.01691	.01729	.01804	.01918	.02089	.02235
.00	.01490	.01422	.01458	.01532	.01646	.01818	.01967

$N = 800$

κ	α						
	2	3	5	7	10	15	20
$-.50$	.02356	.02315	.02315	.02340	.02383	.02449	.02505
$-.40$	.02020	.01979	.01990	.02027	.02086	.02175	.02250
$-.30$	.01701	.01659	.01677	.01721	.01790	.01894	.01983
$-.20$	.01419	.01376	.01397	.01444	.01518	.01630	.01726
$-.10$	.01179	.01134	.01156	.01204	.01279	.01392	.01491
.00	.00981	.00933	.00954	.01000	.01072	.01184	.01282

$N = 1200$

κ	α						
	2	3	5	7	10	15	20
$-.50$	.01812	.01778	.01776	.01793	.01825	.01877	.01921
$-.40$	.01521	.01488	.01493	.01519	.01562	.01629	.01686
$-.30$	.01255	.01222	.01232	.01263	.01312	.01388	.01453
$-.20$	.01026	.00993	.01006	.01038	.01090	.01169	.01238
$-.10$	.00836	.00803	.00816	.00848	.00899	.00978	.01047
.00	.00683	.00648	.00660	.00691	.00740	.00815	.00883

121

Table 2. (*Contd.*)

$N = 1800$

κ	α						
	2	3	5	7	10	15	20
−.50	.01391	.01363	.01360	.01372	.01396	.01435	.01469
−.40	.01143	.01117	.01119	.01137	.01168	.01217	.01260
−.30	.00924	.00898	.00904	.00925	.00960	.01015	.01062
−.20	.00741	.00716	.00723	.00745	.00781	.00837	.00886
−.10	.00593	.00567	.00575	.00597	.00632	.00686	.00734
.00	.00475	.00450	.00456	.00477	.00509	.00561	.00606

$N = 2500$

κ	α						
	2	3	5	7	10	15	20
−.50	.01122	.01098	.01094	.01103	.01122	.01152	.01180
−.40	.00907	.00884	.00885	.00898	.00922	.00960	.00994
−.30	.00721	.00700	.00703	.00719	.00745	.00786	.00823
−.20	.00569	.00548	.00553	.00596	.00596	.00638	.00674
−.10	.00448	.00428	.00433	.00449	.00474	.00514	.00550
.00	.00353	.00334	.00338	.00353	.00377	.00414	.00447

$N = 3500$

κ	α						
	2	3	5	7	10	15	20
−.50	.00899	.00879	.00875	.00882	.00896	.00920	.00942
−.40	.00714	.00696	.00695	.00705	.00723	.00753	.00778
−.30	.00558	.00541	.00543	.00554	.00574	.00605	.00633
−.20	.00434	.00417	.00420	.00432	.00452	.00482	.00510
−.10	.00336	.00321	.00324	.00335	.00353	.00383	.00409
.00	.00261	.00246	.00249	.00259	.00276	.00303	.00327

$N = 5000$

κ	α						
	2	3	5	7	10	15	20
−.50	.00710	.00694	.00690	.00695	.00705	.00724	.00741
−.40	.00554	.00539	.00538	.00545	.00559	.00581	.00600
−'30	.00426	.00412	.00413	.00421	.00435	.00459	.00479
−.20	.00325	.00312	.00314	.00322	.00336	.00359	.00379
−.10	.00247	.00236	.00238	.00245	.00259	.00280	.00298
.00	.00189	.00178	.00180	.00187	.00199	.00218	.00234

Author Index

Subject Index

absolute abundance, 10, 11
abundance model, 3, 28
alpha-diversity, 96

Bayes statistics, 31
beta-distribution of the first kind, 5, 14, 22, 45, 47
beta-distribution of the second kind, 5, 64
beta-diversity, 96
between habitat diversity, 96
biologically controlled diversity, 110
birds, 105
botanical ecology, 29, 102
Brillouin's diversity measure, 10, 74

Chironomid communities, 108
Chi-squared statistic, 97, 100
Christmas counts, 105, 106
class frequency models, 28
class structure, 8
confidence interval, 71, 72, 73
coverage, 67, 86

diversity, 9, 74
Dirichlet distribution, 6, 13, 20, 23, 30, 39, 46

ecology, 3, 85
efficiency of traps, 95
equilibrium distribution, 88, 90
equitability, 74, 76
estimation, 27, 28, 40, 51, 55, 59, 70, 74, 103, 104
exponential distribution, 30, 90
factorial moments, 7
finite populations, 8, 23, 68, 74
Fisher's logarithmic series model, 12, 38, 55, 76

fixed models (populations), 7, 36, 43, 51, 87, 88, 98

gamma-distribution, 4, 6, 38, 45, 64
gamma-diversity, 96
generalized hypergeometric distribution, 7, 47
generalized logarithmic series model, 64
generalized negative binomial model, 63, 98
generating processes, 4, 94
geographic boundaries, 96
geometric series, 21, 47, 55, 79
goodness of fit, 97
Good's indices, 75

higher order classification, 107
hypergeometric distribution, 25

indices of equitability, 74, 76
indices of diversity, 9, 74
infinite populations, 10, 25, 70, 75
information index of diversity, 3, 75, 78, 79, 80, 81

jack-knife method, 56

Lepidoptera, 99
linguistics, 8
logarithmic series distribution, 6, 12, 38, 55, 61, 76, 109
lognormal distribution (model), 5, 7, 27, 57, 80

MacArthur's broken stick model, 30, 63, 66, 76, 79, 86, 90
maximum likelihood, 37, 40, 55, 59, 104